Logical Thinking in the Pyramidal Schema of Concepts: The Logical and Mathematical Elements

Lutz Geldsetzer • Richard L. Schwartz

Logical Thinking in the Pyramidal Schema of Concepts: The Logical and Mathematical Elements

Springer

Lutz Geldsetzer
Universitätsstr. 1
Düsseldorf, Germany

Richard L. Schwartz
New York City
New York, USA

ISBN 978-94-007-5300-6 ISBN 978-94-007-5301-3 (eBook)
DOI 10.1007/978-94-007-5301-3
Springer Dordrecht Heidelberg New York London

Library of Congress Control Number: 2012945628

Printed on acid-free paper

Springer is part of Springer Science+Business Media (www.springer.com)

Photo courtesy of Frank Stiens

IN MEMORY OF

Richard L. Schwartz, Ph. D., J. D.

lawyer, philosopher, gentleman
and a dedicated father and true friend

Richard Leonhart Schwartz was killed in the tornado of New York on July 26, 2012 near his home at Cobble Hill, Brooklyn, N.Y. He was born on April 25, 1951 in San Diego, Cal., USA.

Graduating from Loyola University in Los Angeles in 1973, he continued studies in Philosophy at the University of Duesseldorf, where he got his Ph. D. in Philosophy in 1981. Successively, he enrolled at the Columbia University's Law School and received there a J. D.

After five years' work in a New York law firm, he joined the office of the Attorney General of the State of New York where he worked during 25 years in its Anti-Trust Bureau, lately as its Acting Bureau Chief and Assistant Attorney General.
"A dedicated public servant and attorney, his work ethic and passion were an inspiration to all who had the privilege of knowing him" (G. Cuomo, Governor of the State of New York).

"New York is a better place because of Richard's commitment to fairness and legal excellence" (E. T. Schneiderman, Attorney General of the State of New York).

"Richard knew that he had achieved much of what he considered important in life and was hoping to continue on to complete the rest. Although his plan was thwarted by a lightning strike, he died a happy man" (His brother Dave Schwartz of San Diego, Cal.).

Contents

Translator's Introduction

1 Logic, Empiricism, and Sense Experience

The empiricist, Berkeleyan thesis which underlies this book – that logic and mathematics rest wholly upon sense experience – has found few defenders. The empiricism of the analytic tradition – the dominant movement in philosophy in the English-speaking world for the past century – was from the beginning half-hearted. The founders of analytic philosophy did not ground logic and mathematics in experience. Impressed by the apparent certainty and ideal character of logical and mathematical propositions, they regarded them as the self-evidently necessary structure of successful science.[1] They sought to explain these qualities and the "distinctive function" of logical and mathematical forms while avoiding any Kantian synthetic a priori apparatus.[2]

They asserted the "fundamental significance of mathematics for the construction of the system of knowledge" on the one hand, while insisting on "the purely logical, formal character of mathematics, on which its independence from the contingencies of the actual world depends" on the other.[3] Analytic philosophy thus combined a "latent apriorism" with its widely advertised physicalist empiricism.[4] Logical

[1] R. Carnap's view was that "traditional empiricism rightly emphasized the contribution of the senses, but did not recognize the significance and distinctive function of logical and mathematical forms…", in: *Der Logische Aufbau der Welt (The Logical Construction of the World),* "Preface to the Second Edition", p. XVII, Felix. Verlag, Hamburg, 1998 (originally published 1928). The names "logical empiricism" or "logical positivism" were intended to make clear that a qualified empiricism was intended. See Carnap, "Preface to the Second Edition", p. XVIII.

[2] Wolfgang Stegmüller, *Main Currents in Contemporary German, British and American Philosophy,* 4th German ed. 1969, Engl. translation. Indiana University Press 1970, pp. 264–66.

[3] Carnap, "Preface to the Second Edition" of the *Logical Construction of the World,* p. XVIII.

[4] Geldsetzer, "*Philosophie der Wissenschaften*" (*Philosophy of the Sciences*), § 41 "Logical Empiricism or Analytical Philosophy," unpublished manuscript on file with author; Marc Joseph, *Donald Davidson,* McGill-Queens University Press, 2004, p. 7, noting that "Carnap and the other logical empiricists retain a very powerful apriori apparatus, even if it is not the a priori apparatus envisaged by Kant".

and mathematical propositions – understood since Wittgenstein and Carnap as tautologous and analytic, true by virtue of (conventional) meaning alone – were therefore regarded as incapable of reduction to the empirical, observational level.

Later analytic philosophy is generally understood to have rejected this dualism. Quine, Sellars and Davidson "blur the positivist distinction between … the analytic and the synthetic, the linguistic and the empirical, theory and observation."[5] But in logic, the sharp distinction between syntax and semantics, between uninterpreted, abstract forms and that to which they are applied (models), has remained. Quine proposed a holistic grounding of logic and mathematics in experience, but conceived experience itself behavioristically, as the physical responses of nerve endings, within a context of realism and naturalism. He sacrificed meanings – in technical logical terms intensions, that is, the grasping of a content in consciousness[6] – and embraced the extensionalist thesis that the truth value of sentences is determined solely by their extensions, by which things are being talked about, without regard to intensions.

Geldsetzer rejects the extensionalist thesis, and asserts that no adequate account can be given of logical concepts which does not encompass both their intensional and their extensional aspects. The graphical formalism he proposes for logic (discussed below) is, among other things, intended to render these relationships visible. And as his invocation of Berkeley suggests, Geldsetzer does not conceive sense experience – which he understands as the sole and sufficient basis for logic and mathematics – in a realist and materialist fashion. This is apparent from his doctrine of logical symbols and signs generally. He emphasizes the unity of meaning and physical expression which constitutes a sign, as opposed to the view, dominant in mathematical logic, that signs can be treated as mind-independent physical things – mere sounds or marks, divorced from any meaning at all. Empiricism has frequently gone hand in hand with realism and materialism,[7] but Geldsetzer's empiricism is not of that kind. In his understanding of abstraction and generality, Geldsetzer follows Berkeley. He denies that generality can be achieved by eliminating all sensory qualities; properly conceived, abstraction can only mean that a particular sensory content is treated – inductively – as representative of others of the same kind.[8]

---

[5] Richard Rorty, *Consequences of Pragmatism*, University of Minnesota Press 1982, Introduction p. XVIII.

[6] Quine's rejection of meanings has been most influentially criticized within the analytic tradition by Donald Davidson; see Marc Joseph, *Donald Davidson,* McGill-Queen's University Press 2004, p. 61 and p. 184.

[7] Michael Ayers, "Was Berkeley an Empiricist or a Rationalist", in *The Cambridge Companion to Berkeley*, ed. K. P. Winkler, Cambridge University Press 2005.

[8] Geldsetzer's development of Berkeleyan themes in logic, however, should not be understood to mean that Geldsetzer is in all respects a Berkeleyan. His logic focuses on logical rather than metaphysical questions (indeed, Geldsetzer has not yet developed his own metaphysical views in writing), and a contemporary metaphysics which was "Berkeleyan" in spirit would differ from the version Berkeley himself provided. See A. C. Grayling, *Berkeley: The Central Arguments,* Open Court 1986, pp. 204–210.

In the analytic philosophy of recent decades, empiricism itself is often said to have been rejected or superseded, having already been dealt decisive blows by Sellars, among others.[9] But the empiricism Sellars attacked – the "Myth of the Given" – is not the empiricism of the present book. Sellars understood empiricists to equate the perceptually "given" with mind-independent "non-epistemic facts" which were nevertheless "immediately before the mind." Empiricists, he thought, erred in ignoring the difference between conceptual knowledge and sensation; they treated sensations as though they were logically developed thoughts.[10] And they wrongly regarded the ability to construct logical concepts as a "primordial, non-problematic feature of 'immediate experience.'"[11] Knowledge, Sellars emphasized, involved concepts and language, and should not be confused with the experience of uninterpreted sense data, which empiricists took to be a "self-authenticating" foundation for knowledge.[12]

The empiricism of the present book is not vulnerable to such criticisms. For Geldsetzer, logical concepts are essential to knowledge and can be formed only on the basis of linguistic meanings. In this sense, as discussed below, logical formalization presupposes meaning; it cannot confer sense where none has already been found. There is no appeal to immediately given, uninterpreted sensations. But the approach remains empiricist; there is no role for innate or a priori concepts or capacities; sensory experience is the sole source of our scientific – logically structured – knowledge. Much of our daily discourse and experience is meaningful, but lacks the structure – the unity of intensions and extensions in concepts – which distinguishes logical science.

The preceding remarks are an attempt to locate the present book within a broader philosophical context, but Geldsetzer himself does not address these broader questions here. His focus, as the title suggests, is on the elements of logic and mathematics, and even as to these, the book is not primarily intended as a treatise which explicates and defends theoretical propositions in the philosophy of logic and mathematics. It seeks to analyze and refashion logical elements and provide a usable "how-to" manual for the practical carrying-on of an empiricist logic as an alternative to current logical practice or the lack of it – and some examples (such as the specification of prime numbers in mathematics) of how such a logic may contribute to the development of science. A transformation of fundamental doctrines of

---

9 See, for example, Richard Rorty, "Introduction" to Wilfrid Sellars, *Empiricism & The Philosophy of Mind,* Harvard University Press 1997.

10 "The Myth of the Given is the idea that there can be a kind of *awareness* that ... entails having a certain sort of *knowledge* ...[but] does not presuppose the acquisition of any *concepts...*" (emphasis in original); "Study Guide" by Robert Brandom in: Sellars, *Empiricism and the Philosophy of Mind,* p. 122.

11 Sellars, *Empiricism and the Philosophy of Mind,* p. 59.

12 "[T]raditional empiricism depends on episodes of nonverbal, nonconceptual awareness, which serve both as the raw material for a process of abstraction by which concepts can be formed and grasped and as our warrant for the ground-level (non-inferential) application of those concepts." Robert Brandom, "Study Guide" in Sellars, *Empiricism and the Philosophy of Mind,* p. 168.

philosophical logic is suggested, including new conceptions of logical form and consequence, logical identity, and the nature and function of logical contradictions. Each of these is discussed briefly below.

The objective is to revivify logic by restoring an understanding of its proper relationship to sense experience, so that logic may serve as the unifying methodology of the sciences, our indispensable guide in organizing and expanding our scientific knowledge. Absent such a context, one may share Peirce's concern that logic may become (as it largely has) the preserve of specialists and "in danger of degenerating into a mathematical recreation" on the one hand, while largely vanishing from the practice of many disciplines, on the other.[13]

2 Language, Meaning, Sensory Content, and Logical Form

Risking oversimplification, one can summarize the dominant understanding of logical form as follows: Logical symbols have no meaning as such; they are "uninterpreted" and remain so until a sense is assigned to them by reference to "models", that is, interpretations understood to represent possible states of affairs. The manipulation of such apparently uninterpreted symbols seems possible because logical syntax – the formal rules by which signs can be substituted for one another while preserving truth – is understood to be independent of any relation to possible models or the actual world, a relation termed "semantic".[14] That is why the propositional calculus, which takes sentences as its units, is thought to enable the

[13] "Formal logic must not be too purely formal; it must represent a fact of psychology, or else it is in danger of degenerating into a mathematical recreation." Cited in John Passmore, *A Hundred Years of Philosophy,* 2nd ed. 1966, pp. 142–143.

[14] Generally, the study of language or sign systems (semiotics) is understood to consist of three fields: pragmatics, study of syntax, and semantics. "Investigations in pragmatics are always empirical, since the peculiar character of the speaker can be taken into account only on the basis of empirical determinations. If we abstract from the speaker and limit ourselves to the linguistic expressions and their meanings (or designate) the investigation forms part of semantics. Finally, if abstraction is made also of designate and meanings and attention is confined to the structure of expressions and structural relations between expressions, the study is classified as syntax." As a result: "Syntax makes abstraction not only of the speaker but also of the meanings and references of linguistic expressions. The treatment is purely formal; it is concerned exclusively with the external structure of the expressions. In pure syntax it is again a matter of constructing artificial language systems; but in this case these remain uninterpreted calculi." Wolfgang Stegmüller: *Main Currents in Contemporary German, British, and American Philosophy,* 4th German ed. 1969, Eng. transl. 1970 Indiana University Press, pp. 308–09, 313–14.

Similarly, a useful glossary appended to a standard work in the history of philosophy explains "formal/truth/logic" as follows: "In the traditional use, formal truth means valid independently of the specific subject matter; having a merely logical meaning. In the narrower sense of formal logic which works by exhibiting, often in symbolic notation, the logical form of sentences, it means independent of, without reference to meaning." *Philosophy of Science, Logic and Mathematics in the Twentieth Century,* ed. by S. G. Shanker, London and New York 1996 (Vol. IX of the Routledge History of Philosophy), p. 394.

proof of statements without reference to their meaning. Such proof is believed to rely solely on the (generally assumed) truth or falsity of antecedent statements and the logical syntax, displayed in truth tables, of the connectors which are used to combine them. These formal logical entities and relationships are not considered accessible to sensory perception. Attempts to represent them graphically (for example, Venn's famous diagrams) are treated as helpful visual aids for beginners, not as themselves expressing significant facts about logic, 0.1.3.

This despite the fact that graphical representation of the relations among concepts has important precedents in the history of logic, including "Porphyry's Tree" (Latin: arbor porphyriana).[15] Porphyry's Tree was drawn from the same author's *Isagogue,* an introduction to Aristotle's *Organon* which was influential throughout the development of medieval logic. It shows relationships between generic and more specific concepts, arranged according to their generality, in a tree-like structure which runs from a highest "category" concept such as "substance" at its base to an individual concept such as "Socrates" at the tip of one of its branches. It explicated the process of conceptual analysis and definition through the interplay of the "predicable" terms genus (generic concept), species (specific concept or kind), and differentia specifica (the distinction which 'makes' the distinct kinds within a genus) and was a staple of instruction in logic into the eighteenth century.

The pyramidal logical formalism Geldsetzer proposes is intended to image logical meanings and relationships as directly as the nature of sensorily perceptible symbols allows, 0.1.3., 0.2., 0.10. He denies that logical symbols are properly considered "meaningless" or "uninterpreted", 0.3. That is inconsistent with the unity of meaning and physical expression which constitutes a sign and permits it to function as such. Marks on paper devoid of meaning would not be signs at all, just as, from the other perspective, meaning requires some physical, expressive, substratum for its existence. He introduces his formalism by citing the example of Chinese ideograms – pictorial characters which express ideas through images, 0.10.3.

As noted, an important influence is Berkeley, with his critique of abstract general ideas and his positive thesis that universality is achieved by means of the representative role of particulars. Knowledge and demonstration, Berkeley acknowledged, involve universal notions. What he denied was that this universality was achieved by abstraction from the sensory content of ideas. For, he wrote, "it seems that a word becomes general by being made the sign, not of an abstract general idea, but of several particular ideas, any one of which it indifferently suggests to the mind" so that "an idea, which considered in itself is particular, becomes general, by being made to represent or stand for all other particular ideas of the same sort."[16] So, for example, a particular black line an inch in length may be drawn by a geometrician upon a piece of paper, and "as that *particular line* become general by being made a

---

[15] W. and M. Kneale, *The Development of Logic*, 1968, p. 232; *Historisches Wörterbuch der Philosophie,* Art. "Arbor porphyriana, porphyrianischer Baum", vol. 1, p. 493, 1971; *Enzyklopädie Philosophie und Wissenschaftstheorie,* Art. "Arbor porphyriania", vol. 1, p. 152, 2004.

[16] "The Principles of Human Knowledge", Sec. 11–12, in *Berkeley's Philosophical Writings,* ed. by David Armstrong, 1965, pp. 50–51.

sign, so the *name* line, which taken absolutely is particular, by being made a sign, is made general. And as the former owes its generality, not to being the sign of an abstract or general line, but of all particular right lines that may possibly exist, so the latter may be thought to derive its generality from the same cause, namely the various particular lines which it indifferently denotes."

In short, Berkeley's theme is that while we certainly distinguish, within perceptions, among different sensory contents, we cannot dispense with these in framing concepts: "[A] man may *consider* a figure merely as triangular, without attending to the particular qualities of the angles, or relations of the sides. *So far he may abstract.* But this will never prove that he can frame an abstract, general, inconsistent *idea* of a triangle."[17]

In the development of modern philosophy of logic, however, it has been not Berkeley but Frege, with what was understood as his successful criticism of psychologism, who has been influential. And what Berkeley affirmed – the sensory content of all concepts – Frege denied.[18] Numbers, he insists, have nothing to do with the subjective sense impressions or mental images which different individuals may arbitrarily associate with them. These latter, he argues, are as irrelevant to mathematical thought as such as are the chalk marks by means of which mathematical symbols are written on a blackboard. It is an error, Frege asserts, to confuse the mental and physical conditions of the fact that a proposition is thought with a proof of the truth of that proposition. For a proposition, he says, does not cease to be true when I am no longer thinking of it, any more than the sun ceases to shine when I close my eyes.

Frege therefore drew a sharp distinction between the psychological and the subjective on the one hand, and the logical and the objective on the other.[19] To mark

---

[17] Sec. 16, p. 54 (emphasis in original).

[18] Gottlob Frege, *Die Grundlagen der Arithmetik: Eine logisch mathematische Untersuchung über den Begriff der Zahl,* Breslau, 1884, Einleitung (Introduction), pp. 5–7. In his *Begriffsschrift* (eng. *Conceptual Notation*), Frege believed himself to have shown how "pure thought, irrespective of any content given by the senses or even by an intuition a priori, can, solely from the content that results from its own constitution, bring forth judgments that at first sight appear to be possible only on the basis of some [sensory] intuition." *Conceptual Notation and Related Articles*, transl. and ed. by T. Bynum, Oxford 1972, p. 167, quoted in H. W. Noonan, *Frege: A Critical Introduction*, 2001, p. 21.

[19] In this, historians of philosophy have pointed out, he was anticipated by Bernard Bolzano, who distinguished between "subjective" and "objective" representations; the latter "do not require a subject but subsists … not indeed as something *existing* but as a certain *something* even though no thinking being may ever have it…" B. Bolzano, *Theory of Science* (*Wissenschaftstheorie*), transl. and ed. by R. George, University of California Press, 1972. Coffa has given this distinction a place of decisive importance in his account of the "semantic tradition": "Objective representations are the substance (*Stoff*) or content of subjective representations. Their being in no way depends on the existence of subjective acts, just as the meaningfulness of expressions in no way depends on anybody's bearing the appropriate meaning in mind; and like meanings there is only one for each linguistic unit unless the given expression is ambiguous. Clearly, Bolzano's objective representations are the 'meanings' or 'senses' of his successors in the semantic tradition. The distinction between objective and subjective representations amounts to a separation of meaning from psychological processes." J. Alberto Coffa, *The Semantic Tradition from Kant to Carnap: To the Vienna Station,* Cambridge University Press 1991, p. 30.

the divide, he tells us, he uses the term "representation" (German: "Vorstellung") in a purely psychological sense, which he distinguished from concepts and objects. Linked to this is his famous injunction (known as the "context principle") to seek the meaning of a word in its objective propositional context, thus understood, rather than in isolation from it; for the latter approach, be believed, would compel us to designate internal images, understood as the content of individual minds, as the meanings of words.

The advances in the "formalization" of logic with which Frege is now generally credited presuppose this distinction between the subjective and the logical as well as the separation of syntax from meaning. Formalization in this sense "consists in ignoring every *meaning* of the signs employed and considering the signs exclusively according to their written form."[20] We then operate with these signs solely on the basis of certain "transformation-rules concerned only with the written shape of the signs." (at p. 40). This was the view which was made popular in the English-speaking world through Ayer, who emphasized that logic was not to be thought of as concerned with "the workings of thought … What it was actually concerned with was the formal relationship of classes, as shown by the fact that all of its principles of inference are subsumed in the Boolean class calculus, which is subsumed in its turn in the propositional calculus of Russell and Whitehead."[21]

Ayer linked this view to the doctrine of analyticity, that the truths of logic and mathematics are "analytic" in the sense that their validity follows simply from the definitions of the terms they contain. We could not abandon them, he argued, without contradicting ourselves, without "sinning against the rules which govern the use of language" (p. 77). The analytic character of logic and mathematics had not been seen earlier because logic had been mixed with psychology, so that traditional logic had been "insufficiently formalized" (p. 80–81).

This is the conception of formalization which emerges from the logical tradition of Leibniz, Boole, Frege and Russell, and which still underpins the propositional calculus and modern mathematical logic generally. A strict separation of meaning (semantics) from form (syntactics) and the extensionalist thesis are assumed, and the result is that strings of symbols are treated like equations and substituted for one another without reference to their meaning (intension), provided only that they are "true" in an extensional sense, that is refer to the same things. The model for this view of logic is mathematical calculation itself; where, it seems, one can perform

---

[20] I. M. Bochenski, *The Methods of Contemporary Thought (Die Zeitgenössischen Denkmethoden)*, Eng. trans. by Peter Caws, Harper & Row, 1968, p. 34–35. Thus, "the word with which semiotics is concerned is the material word, that is, in the case of the spoken word, a set of waves in the air, and in the case of the written word, a series of ink-marks on paper. That the word has to be taken in this sense is already clear from the fact that it is set over against what it means."

[21] *Language, Truth, and Logic*, 2nd ed. 1946, p. 80–81. Ayer summed up his view as follows: "[F]ormal logic is not concerned with the properties of men's minds, much less with the properties of material objects, but simply with the possibility of combining propositions by means of logical particles into analytic propositions, and with studying the formal relationships of these analytic propositions."

operations correctly – that is, calculate – without knowing why one procedes in the way one does, if one simply follows the relevant syntactic rules.

Geldsetzer uses the term "formalism" in a broader sense. The type of formalism adopted is in fact the most important decision in logic, for it is this which determines how thoughts are to be given a logical form, 0.1. But he regards mathematical logic as only one among several types of formalisms which the history of logic exhibits. He agrees that the dominant model for the formalism in mathematical logic is arithmetical calculation itself, so that in this tradition logic is thought of as a "calculus", a tool for calculating, 0.1.2. But there are other possibilities, other models for giving logical form to thoughts. The most prevalent in the history of logic, he points out, is the model of phonetically written languages, where letters evoke sounds, which in turn evoke meanings. This is the kind of formalism which underlies conceptions of logic as an ideal or artificial language, 0.1.1.

In this connection, Geldsetzer offers both a critical thesis and a positive proposal. His criticism is directed to the notion of formalization fundamental to mathematical logic: The radical separation of syntactic form from meaning is infeasible; logical consequence – that is, the truth-preserving or "valid" relation between premises and a conclusion – must always remain situated within a thematic context of meanings. Geldsetzer's positive proposal is a formalism based upon direct graphic representation of logical elements, akin to the mode of representation of Chinese pictorial characters, which directly represent meanings through images. As a "remembered schema" which enters into logical thinking, "graphical formalisms become an image of the essence of logic", 0.2.

If logical symbols presuppose meanings and sensory content, from what sources are those meanings drawn? If we examine our actual understanding of the customary logical symbols we will see that the logical use of these letters and signs depends on an understanding of meanings derived from their use in language (spoken or written) or mathematics, 0.6.1. Such symbols, when employed for logical purposes, acquire supplementary, specifically logical meanings; but our prior acquaintance with them in the context of language or mathematics supplies a necessary substratum on which logical meanings can be layered. Letters, for example, are used and understood in language as signs for certain sounds; and we presuppose this meaning of letter and incorporate it when we conceive of letters in logic as names of concepts.

That language is prior to logic in this sense does not mean that logic can be reduced to language. The relationship is one of dependence in the context of further development. On the one hand, logic must begin from and cannot itself supply an understanding of meaning as captured in a linguistic articulation. The process of logical formalization presupposes that a statement or other linguistic articulation has been interpreted; it cannot confer sense where none has been found. On the other hand, logical formalization cannot dispense with meanings, but remains tethered to them. Meanings remain at each step a necessary underpinning of logical thought, irrespective of how much further they may be logically formed and developed.

Again, this is not to say that logic is simply an extension of language. The foundational layer of meaning which language supplies is essential, but logical thought is an achievement by no means invariably present in language (although historically logic has reacted back upon it, molding language according to logical forms), 0.10.2. Our grasp of a particular meaning is what Geldsetzer refers to as an "intension". This is what occurs when we understand a word. But such purely linguistic meaning can be (when undisciplined by logic) and often is, in our understanding of language and literature, somewhat free-floating. Logical concepts, in contrast, must always consist not only of sensorily grounded ideas, or intensions, but of extensions, which fix their range of application. The task of logic is to unify thought and give it a structure, in accordance with a formalism, which language alone does not provide, 0.3.1., 0.4. The core of the formalism Geldsetzer proposes is the fixing of intensions which we grasp at different levels of generality; but these are never unconnected to at least one sensory instance – an extension. "Ordinary language" is therefore in no position to act as an incorrigible authority regarding logical relationships; it serves instead as one source of meaningful elements which have been adapted in logical formalisms for logical use.

3 Intensions, Extensions, and Concepts

The logical tradition since Aristotle has formalized concepts as units, suggesting that they are indivisible entities, something like the atoms of thought. But concepts have also often been thought to have dual aspects – roughly the content or meaning and the extent or range of application of the concept – which have been expressed in varying ways. Content is often said to consist in the characteristics or features which we "have in mind" when applying the concept; the extent or application is said to be the set of things which "fall under" the concept or which it is "true of". In the twentieth century, the terms "intension" and "extension", introduced most notably by Carnap, came into widespread use in mathematical logic. But there the usual starting point for analysis has been not the concept but the expression – the word or sign taken as a physically existing object, as sound or marks on paper, as distinguished from its meaning. Consequently, it is expressions, not concepts, which are often said to have both extensions and intensions. But for Geldsetzer, extensions and intensions are aspects of concepts.

Geldsetzer takes this duality as fundamental, as the key to what concepts are. He emphasizes that the intensional and extensional aspects of concepts must be clearly distinguished, treated as separate elements in logic and our mental life generally, and separately formalized. Our ordinary language and day-to-day mental functioning are distinguished from logical thought precisely by the fact that their content consists of intensions – more or less determinate ideas – without any systematic attention to extensions. It is not concepts, but intensions which are generally designated by the words of phrases of a language.[22] Ordinary language is thus usually meaningful,

[22] L. Geldsetzer, *Grundriss der Pyramidalen Logik mit einer logischen Kritik der mathematischen Logik und Bibliographie der Logik* (Internet HHU Duesseldorf 2000) p. 20.

but not, in the strict sense, conceptual or logical. We form concepts only when we supplement meaning with extensions, by fixing which phenomena "fall under" a particular idea. Only then are we thinking logically, for logic consists essentially in conceptual thought, in which intensions must be continually tested against extensions – against our sensory experience – , and the results reviewed to determine how well they fit with other concepts and theories.

There can be no genuine logic which does not take account of – and no perspicuous logical formalism which does not plainly show – both the intensional and the extensional aspects of the concept. Judged by this standard, some traditional and current theories of what logical concepts are and the manner in which they are formalized are deficient or one-sided.

Extensionalists such as Quine reject intensions as without foundation in reality. They are often motivated by a robust, common-sense, realism – expressions are the names of independently existing things, which remain unaffected by how we conceive them.[23] It was this view of expressions as names which Carnap rejected in *Meaning and Necessity*. Previous analyses of meaning, he thought, had all "regard[ed] an expression in a language as a name of a concrete or abstract entity. In contradistinction, the method here proposed takes an expression, not as naming anything, but as possessing an intension and an extension."[24]

Intensions are "ideas" in our modern, psychological understanding of that term; the representations which arise when we understand a word. It is intensions or meanings, and not concepts as such, which are, as noted above, conveyed in our every-day communication with each other, and which form the object of dictionary definitions, the study of semantic aspects of language, and of the history of ideas.[25]

Intensions, then, can be and usually are grasped in mental acts and designated by words without any close connection to their proper extensions; but extensions can only be grasped through intensions. Extensions are those factual particulars in which something common can be found and retained as an intension, so that they can only be designated through complete concepts, not through words as such. The relations of intensions and extensions in concepts are inversely proportional: The larger the number of intensions incorporated in a concept, the more limited its extension, and conversely, the smaller the number of intensions, the broader the extension of the concept becomes. These relationships are made visible in the pyramidal formalism proposed here.

Extensions are there shown through the subordination of conceptual positions, and the pyramid shows as well that logical quantification presupposes a conceptual

---

[23] "It is an affront to common sense to see a true sentence go false when a singular term in it is supplanted by another that names the same thing … What is true of a thing is true of it, surely, under any name." Quine, *From Stimulus to Science*, Harvard University Press 1995, pp. 90–91, cited in Gary Kemp, *Quine: A Guide for the Perplexed,* Continuum 2006, p. 129.

[24] "Preface", iii, *Meaning and Necessity,* University of Chicago Press, 2nd ed. 1955.

[25] See Geldsetzer, "Wörter, Ideen und Begriffe: Einige Überlegungen zur Lexikographie", in: C. Strosetzki (ed.) *Literaturwissenschaft als Begriffsgeschichte,* Felix Meiner Verlag 2010, pp. 69–96.

position consisting of both intensions and extensions. For one must start from a particular conceptual position in order to say whether, with respect to other conceptual positions, all, or some, or a single one, or none of them (not one, not any) fall under it.[26]

4 The Pyramidal Formalism: Intensions and Extensions

In Geldsetzer's pyramidal formalism, capital letters designate intensions, defined as recollected representations of sensory experience, 1.1. This is a departure from a logical usage traditional since Aristotle, and largely continued in modern mathematical logic, in which letters designate entire concepts. Geldsetzer conceives concepts not as atomic units but as structured relations between intensions and extensions, the latter defined as representations which circumscribe the range of experience to which the concept is applicable, 1.2. In the pyramid, the extension of a concept is shown by its position vis-à-vis lower level concepts. Both components are essential; a genuine concept cannot dispense with either an ideational content or some range of application – though this latter need comprise no more than a single instance.

The order of the concepts among themselves flows from their structure: "Concepts relating to a particular field of experience group themselves by means of their intensional and extensional interlocking structures within the pyramidal formalism", 1.4. The order is of declining – or ascending – degrees of generality; from top to bottom, the relationship runs from categories and generic concepts, through species and subspecies, to lowest-level concepts, or individuals. "Higher" and "lower" do not imply any valuation; the pyramid might equally well be reversed, and represented as a tree with spreading branches so that the order of more general and more specific concepts would be reversed. One can bring out what is "logical" in this formalism by contrasting it with a purely extensional classification system. Such a classification system might take an inventory of objects based solely on their presence in a particular space, and without the internal structure of intensions and extensions which gives shape and internal order to the formalism, Geldsetzer has put forward. But a logical system is animated by the fixing of identical, sensorily grounded intensions at different levels of conceptual generality. Logical concepts, in short, always consist of both intensions and extensions.

Concepts and pyramids themselves are formed by induction. Through a comparison of lower level concepts – sensory instances – common characteristics are

[26] Geldsetzer explains mathematical quantification as a special, more precise form of logical quantification, such that the logical unit "a" (single) serves simultaneously as the unit of measurement, the logical "all" becomes the mathematical infinite, and the logical "none" (not one, not any) becomes the "zero" measurement. It is on this basis that Geldsetzer constructs a logical foundation for the concept of number and arithmetic. See *Grundriss der Pyramidalen Logik* (Internet HHU Düsseldorf 2000) p. 23, and the present book n. 1.16.

separated from specific differences. Characteristics which are common to multiple concepts – and are therefore identical – are designated by the same letters and are separated to form higher level generic concepts, 1.4. Logical identity can properly be attributed only to intensions which are represented by the same letters. These reappear in concepts at multiple levels of the pyramid. For example, the intension "living" is a characteristic of all of the distinct concepts which fall under it at varying levels of generality, such as "animal", "mammal", "dog" and even the singular dog named "Sofie". Each of these conceptual structures consists of a somewhat different cluster of characteristics – specific differences distinguish each from the generic concept above it and the coordinate concepts at the same level; mammals, for example, unlike other animals, bear their young alive – but in each such concept "living" will be designated, when formalized, by the same letter. See Figures 1 and 2 at 1.4.

5 Logical Identity, Sense, and Reference

Accordingly, certain of the intensions which constitute distinct concepts are identical, in a strict sense, but distinct concepts never are. That which is identical is the same, and is to be distinguished from that which is equal to, or like, for these latter are predicates which properly apply to two or more distinct logical structures, which are "equal" or "like" in some but not in all respects, and therefore distinct.[27]

The sense of "identity" Geldsetzer employs here reaches back to Aristotle and was well expressed by Locke, who wrote that "the foundation of all our knowledge lies in the faculty we have of perceiving the same idea to be the same and of discerning it from those that are different."[28] Thought presupposes identity in this sense; identity is what characterizes the grasp of a particular meaning. As we shall see below, Geldsetzer criticizes the usual logical formalization of identity, $x = x$, as improper because self-reflexive. In the formalism he proposes, identity is expressed by using the same sign for the same meaning; identity might be formalized as simply "x". In Geldsetzer's pyramidal formalism, as we have seen, identity is expressed by the use of the same sign for the same meaning.

In modern mathematical logic, however, logical identity may be represented not by the same or like signs but also by different signs, for a very different conception of logical identity is at work. Frege provided its standard formulation with his distinction between sense and reference.[29] Sense is the way in which the sign (or an object) presents itself to us – how it designates – and its reference is that

[27] Geldsetzer, *Logik,* Aalen 1987, pp. 76–78.

[28] *An Essay Concerning Human Understanding,* ed. A. C. Fraser, 1st ed. 1891, reprinted 1959, 2 vols., vol. 2, Book IV, Chapter 8, p. 294.

[29] "Ueber Sinn und Bedeutung" ("Concerning Sense and Reference"), originally in *Zeitschrift für Philosophie und Philosophische Kritik,* 1892, pp. 25–50, cited according to G. Frege, *Funktion, Begriff, Bedeutung,* ed. by M. Textor (2002), pp. 23–46.

which it designates, usually understood as a mind-independent object or state of affairs. For Frege, the distinction provided the answer to the question of how contingent statements of identity could be true and yet informative, since we might not initially understand them to express identity at all. He illustrated this with his famous example of the different "senses" or appearances to us of the morning star and the evening star, each of which was later discovered to "refer" to the same planet, Venus. As the example suggests, for Frege truth values depend on reference, and in that sense all true statements refer to something identical, as do all false statements. The logically identical is something which lies behind logical symbols and does not appear in them – an expression of the rationalist separation of logical thought and sense experience which Geldsetzer rejects.

Geldsetzer finds Frege's distinction between sense and reference "hermeneutically deficient", 3.1.1. "From the hermeneutical standpoint, the correct formulation is that signs possess a double meaning. One of these meanings (Frege's 'sense') varies according to the formalism which is employed; for example, the sound value of letters, the numerical value of symbols for numbers. The other meaning is that which is properly termed logical or mathematical (Frege's 'reference'). When the formalism is applied to actual states of affairs, the (linguistic) meaning of the concepts and expressions takes the place of the logical or mathematical meaning."

Geldsetzer denies, then, any realistically conceived "reference" behind the signs as the locus of truth and identity. A very influential stream of thought represented by Saul Kripke has pushed in precisely the opposite direction. In his *Naming and Necessity*,[30] Kripke, in the context of a discussion of proper names, rejected the "descriptivist" thesis of Frege and Russell that these necessarily had meaning or sense, although he allowed that such descriptive meanings might serve to "fix the reference" of a name.[31] What Kripke insists on is that no set of meanings or descriptions – Frege's "sense" – can be synonymous with a proper name, for, he argues, error and ignorance as to the bearer of the name are perfectly compatible with "successful reference". Aristotle, for example, might not have been the teacher of Alexander the Great, but that does not prevent us from thereby successfully referring to Aristotle, rather than to some other Greek philosopher of antiquity. The tendency of Kripke and his followers is therefore to reject meanings in favor of a theory of direct reference; that is, to contend that the propositional contribution of a proper name like "Aristotle" is not a sense – a meaning or description – but its referent.[32] But the current status of the discussion appears to be that even many of those who agree with Kripke's criticisms of descriptivism have not been satisfied that any satisfactory alternative account of meaning and reference has been given.[33] This is perhaps a sign that a fundamentally different approach is indicated, and this Geldsetzer offers.

---

30 1972, 2nd ed. 1980, Harvard University Press.

31 *Naming and Necessity,* p. 59.

32 See the useful account of Kripke and direct reference theory given in *Logic: Key Concepts in Philosophy,* by L. Goldstein, A. Brennan, M. Deutsch and J. Y. F. Lau, 2005.

33 *Logic: Key Concepts in Philosophy,* pp. 109–10.

From Geldsetzer's standpoint, the nature of the relationships among logical symbols must be rethought. What are the possible and proper ways in which logical symbols may be set in relation to each other? Much of the present book may be thought of as an answer to this question. To introduce that answer, I next outline briefly Geldsetzer's treatment of logical connectors in general, and of the equivalence connector in particular.

6 Logical Connectors

In modern propositional logic logical connectors are defined operationally, by means of truth tables. These are themselves diagrammatic summaries of rules of logical syntax (the justifications for which are far from clear) stating which sequences and combinations of true and false statements constitute valid, that is, truth-preserving, inferences. Logical connectors are not, in themselves, thought to represent anything. As Wittgenstein put it: "My fundamental thought is that the 'logical constants' do not represent."[34] In classical logic since Aristotle a similar view dominated; logical connectors were thought to be "meaningless" in themselves and to acquire logical significance only in conjunction with expressions and propositions, 2.

Geldsetzer rejects this view, and argues that logical connectors are best understood as drawn from ordinary language and as retaining and presupposing the meaning they have as grammatical elements.[35] They acquire logical meanings from the relations among conceptual positions in the pyramid: "Logical connectors are logical concepts for the description and interpretation of the relationships which obtain among the conceptual positions of the pyramid", 2. As concepts, they have distinct meanings and scopes of application and can therefore themselves be understood as forming a pyramid. See diagram at 2.2. They are symbolized by arrows, which show the direction(s) in which the various connections they express properly run. When the connectors are used to form propositions, the truth or falsity of the latter hinges on whether the connection runs in the direction which is specific and proper to the logical connector employed. Connections within the pyramid may run in either of two possible directions, vertical or horizontal, or in both, 2.3.

At this point Geldsetzer introduces an important type-distinction among logical connectors. Only some are capable of binding concepts together into propositions with truth values. Others result in non-propositional expressions which are incapable of truth or falsity. The underlying distinction – between propositional assertions, which may alone be true or false, and expressions, which may not – is one which has been obscured in logic since Aristotle.[36] Geldsetzer's fundamental contention is

[34] *Tractatus Logico-Philosophicus*, 4.0312, cited according to *Ludwig Wittgenstein Werkausgabe* (Suhrkamp) Frankfurt am Main 1984, vol. 1, p. 29.

[35] Geldsetzer, *Grundriss der Pyramidalen Logik*, (Internet HHU Duesseldorf 2000), p. 56.

[36] *Grundriss der Pyramidalen Logik* (Internet HHU Duesseldorf 2000) p. 55.

simple: An expression, such as "Jim or John" is not capable of being true or false. Only an assertion – for example, "Jim and John are brothers" can be so. But the distinction has not been sufficiently taken account of because alternative and conjunctive expressions, instead of being separately attended to, were subsumed into the classical distinction of propositions as alternative and conjunctive.

The same view has dominated in modern propositional logic, where letters generally stand for propositions rather than concepts. There, an alternative (p or not-p) is usually regarded as true if at least one of its branches is true, while a conjunction (p and q) is true only if both of its conjuncts are true. The mischief in this tendency is the suggestion of "higher" or meta-levels of truth and falsity than those which enter into the expression itself. One quickly moves from the fact that "p and not-p" is a mixture of truth and falsity to the supposed meta-level, where the contradiction as a whole is understood to be necessarily false.[37] Since Russell, these meta-levels have been the favorite logical means of avoiding contradiction and paradox. Geldsetzer's thesis, as we shall see, is that any such attempt at avoidance fails.

Vertically, Geldsetzer interprets the relation of implication between one concept and another as general (aristotelian) attribution which connects in both directions. It signals both the inherence of generic characteristics as intensional attributes in the specific concepts which fall under that generic concept (usually called "material implication" or subsumption), and the extensional subordination of such lower-level concepts to a generic, higher-level one (usually called "formal implication" or inclusion), 2.4.1 Horizontal implication between concepts occurs by means of the correlative implication, which may be read as "if one, than the other", 2.4.2. It serves as the logical formulation of both correlative and causal relationships. The "causal" implication requires for its proper application a generic concept which is common to the correlated concepts, and which contains their common generic characteristics, that which is identical to each of them. The specific differences of the correlated concepts express that which distinguishes them, 2.4.3. The non-causal correlative implication may also combine "horizontal" concepts absent such common generic characteristics.

Both the vertical and horizontal connectors comprise more specific connectors which are mutually exclusive. The specific connectors which fall under the correlating connector are the existential connector ("there is A") and its counterpart, the negation of the copula ("AB is not AC"), 2.4.5. Negation occurs only between coordinate concepts. "The existential connector cancels [this] distinction between coordinate concepts and converts their common (generic) characteristics into a higher level concept." The equivalence connector is formed by fusing together these mutually exclusive specific connectors into a single connecting concept. "Through this fusion they reciprocally cancel their respective functions, which generated truth values", 2.7. That is why the equivalence is capable only of forming expressions, rather than propositions with truth values, 2.10.1.

[37] *Grundriss der Pyramidalen Logik*, p. 56.

7 Equivalence and Singular and Particular Propositions

Geldsetzer regards equivalence as in general expressing relations of synonymy; it is the relation which obtains, for example, between the different parts of an utterance (taken together) and its unifying meaning. The two sides of the equivalence (symbolized by the equals sign) are not the same, and must therefore be distinguished; but at the same time, the existential connector introduces both of the distinguished sides as a single concept, 2.10. Therein lies a contradiction.

It is this meaning which should be assigned to the customary use of the equivalence connector (equals sign) in mathematical logic and mathematics – although this is not usually recognized. In that context it signifies a relation between logical or mathematical structures which are distinct but which may be substituted for each other because they have the same meaning. Geldsetzer's thesis is that, so understood – as expressing the unity but not the difference of the two sides – this notion of equivalence explains how this connector functions in important logical contexts. These include singular and particular propositions and definitions generally.

Geldsetzer's argument begins with an assessment of Kant's classical distinction between analytic and synthetic propositions.[38] He approves Kant's account of the nature of analytic judgments as expressing generic characteristics which inhere in the concepts we use. Here the material implication between concepts, which is also properly expressed by the copulative "is", finds its proper place ("roses are flowers"). But Kant erred in his account of synthetic judgments. These he characterized as attribution of a predicate ("red") to a general concept such as "rose", not belonging to its intensions.[39]

Kant's error lay in his misunderstanding of how "synthetic" predicates are attributed to concepts. This occurs through equivalences ("that means" or "="), never through genuinely copulative judgments ("is"). He chose a mathematical equation as an example of a "synthetic judgment (a priori)", but simultaneously erred in characterizing such equations as copulative judgments – an error in which most mathematicians and logicians have followed him ever since.

With respect to "synthetic judgments (a posteriori)", Kant attempted to provide an example taking the form of a regular copulative judgment expressing the synthesis of "rose" and "red". Here again, Geldsetzer believes, he went astray. Being "colored" is certainly a specific characteristic of the concept "rose", for a rose must be of one color or another. So one can properly and analytically assert: "(All) roses are colored". But that does not mean that "red" is a proper synthetic determination of the concept "rose". The logical function of "red" is rather to identify more precisely a *kind of* or *an individual* rose which was only gestured at by means of the phrase

---

[38] L. Geldsetzer, *Grundriss der Pyramidalen Logik,* p. 60.

[39] Kant, *Kritik der reinen Vernunft,,* (Felix Meiner Verlag) Hamburg, 1976, "Von dem Unterschied analytischer und synthetischer Urteile", A 7, B 11, p. 45; English transl. by Norman Kemp Smith, *Immanuel Kant's Critique of Pure Reason*, St. Martin's Press, New York City, 1965: "The Distinction between Analytic and Synthetic Judgments", p. 48.

"some roses" or "this rose". It is an equivalence, rather than an assertion, which is thereby constructed, and it reads "some roses = red roses" or "this rose = a red rose". It brings into relation two expressions with a meaning which is common to both but with the distinctive characteristics that the former expression is of a quantified, indeterminate concept, the latter more specifically formulated. Put another way, an incompletely specified concept ("some roses" or "this rose") is defined by means of one which is more precisely specified ("some red roses" or "a red rose").

Equivalence also exists between definitional expressions such as "some living things" and "animals", so that we may properly write "some living things = animals" but also, with equal correctness, "some living things = plants". Here again we have the same structure. A meaning which is common to two different expressions on different sides of the equivalence sign, one of which, however, has a more specific, differentiating character than the other, signaling that a more particular species concept already hinted at ("*some* living things") is being given more precise expression.

This reflects a new understanding of the logical structure of singular and particular statements. Since Aristotle, these have usually been understood as true or false assertions expressing attributions of predicates to concepts or to particular things. Geldsetzer's contention is that particular and singular statements, through which we construct "concrete" objects, are properly interpreted not as assertions but as equivalences, which express both identity and difference, and therefore are not properly true or false. "A rose is a flower" is analytically true, for it expresses that the concept "flower" is contained as an intension in the concept "rose"; "A rose is an animal" is analytically false, for the concept "animal" is not so contained. But the supposedly "synthetic judgments" "a rose is red" or "some roses are red" is strictly speaking both true and false because there are some roses and individual roses which are "not red". Accordingly, such so-called synthetic judgments are more properly understood as equivalences defining a vaguely identified individual ("this rose = a red rose") or a specific kind ("some roses = red roses") by means of more detailed characteristics, thus combining different expressions of specified concepts with the same meaning.

Geldsetzer does not discuss the epistemological and ontological implications of this doctrine, but one sees that they accord with empiricism and idealism. Ontologically, it suggests a departure from the Aristotelian model of mind-independent particular things with respect to which qualities are predicated. Epistemologically, the distinction between analytic and synthetic has generally been understood as dividing knowledge gained through examination of concepts or meanings from knowledge gained by experience. Geldsetzer's empiricist program relativizes that distinction; logical concepts themselves are and must be drawn from sense experience. The opposite side of that coin is that new experience itself is not non-conceptual. At least to the extent we articulate it logically, new experience is a matter of forming more particular concepts – the "red rose", as opposed to simply "rose" or "that rose" – which stand in a relation of equivalence to those we already possess.

This understanding of equivalence also sheds new light on old questions as to how unity and multiplicity can be reconciled in logical thought. Zeno argued in defense of Parmenides that the existence of multiplicity presupposed that the same things were both like and unlike; Zeno regarded this latter proposition as a contradiction and therefore believed himself to have successfully refuted the existence of the many by means of a *reductio ad absurdum*.[40] Geldsetzer, as we will see below, embraces the contradiction and in the equivalence has provided an answer to Zeno showing how concepts can indeed be both like and unlike.

8 Definitions and Equations

Definitions and equations are therefore equivalences, not genuine assertions or copulative judgments, which can be unequivocally true or false. In fact, if equivalences are misunderstood as assertions or copulative judgments – which they usually are – they would combine unity and difference in contradictory fashion.

Our traditional understanding of definitions supports this view. We speak of terms being defined more or less correctly, but not of the truth or falsity of definitions. And there is another indication that definitions are not properly considered as assertions. Generally, relying on the principle of bivalence or the excluded middle (a meaningful assertion is either true or false – there is no third alternative) we expect that negating a general statement will convert its truth value from true to false, or vice versa. That is not the case for definitions, which do not exclude their negations, but are consistent with them. For example, "some living things = animals" is consistent with "some living things = plants" although plants are not animals, and animals are not plants. If the terms of a definition are negated, the result is not a false statement, but concepts defined by means of negation. For example, "some living things = non-animals/some living things = non-plants", 3.5., 3.5.1., 3.5.3.

Geldsetzer's characterization of equations as equivalences is a departure from the dominant view in the philosophy of mathematics of equations as assertions. For Kant, correct equations were synthetic a priori truths – that is, propositions which gave us information about the world which we could know to be true prior to experience. Many subsequent philosophers, most notably Wittgenstein and the Logical Positivists, have regarded equations as analytically, that is, tautologously true propositions. They were right to understand equations as tautologous, but they erred doubly in maintaining that equations are assertive propositions to begin with (instead of equivalences) and that logical propositions are generally tautologous. In contrast, Geldsetzer's view is that equations are definitional expressions showing equivalence between, for example, different numerical expressions. In the equation $5+7=12$ what the two sides of the equation have in common is their numerical value; what differs are the means of expressing that value. As Geldsetzer explains: "'x = y' does not assert that x and y are identical, but that they have, as different

[40] Plato, *Parmenides*, 127 d–e, transl. by F. M. Cornford, in: *The Collected Dialogues of Plato*, ed. by E. Hamilton and H. Cairns, Princeton University Press, Princeton, New Jersey, 1961, p. 922.

signs, the same meaning. And in arithmetic that meaning can be any numerical value, under the condition that that value is the same for x and y." Further, what is often called a "false" or "incorrect" equation (for example, $3+2=7$) "is therefore not a false assertion, but something which from a logical point of view may be termed a conceptual confusion; as if one were to say, 'dogs and cats, that is birds'".[41]

9 Reflexive Self-Identity

Geldsetzer is deeply critical of the usual logical formulation of identity ("$x=x$"). Before summarizing that criticism, however, it is important to specify the sense of identity which in his view is presupposed by logic and thought generally. This is identity in the sense in which Locke spoke of it as the faculty of perceiving the same idea to be the same – in Geldsetzer's terms, the grasping of a meaning, or intension.[42] As Aristotle wrote in the *Metaphysics:* "Suppose 'man' has the meaning 'two-footed animal'. By 'having a meaning' I mean this: if 'man' is 'two-footed animal', then if anything is a man, its 'being two-footed' will be what its 'being-a-man' is… It is not possible, then, for being a man to mean what 'not being a man' means, for 'man' … has a definite meaning."[43] This is identity as a notion, which refers to the determinateness of the meanings – as ideas or intensions – we grasp in perception and thought.

The logical formulation of identity which Geldsetzer criticizes is not transitive but self-reflexive, formalized as "$x=x$". This is understood to mean that x – that is, something, a concept or an object – is identical with itself. Thus, it is often said that identity is a "relation that every object bears to itself and to nothing else."[44] Many since Leibniz and Frege have regarded this as a logical axiom, and yet doubts concerning the utility of such a principle are not new. Locke, while regarding such "identical propositions", in which a term is predicated of itself, as "true and self-evident", thought that they were empty and could teach us nothing.[45] Nevertheless this understanding of identity continues to have influential defenders. Notable among them is Kripke, who takes as given the premise that "every object is surely necessarily self-identical" and denies that this leads to paradoxical consequences.[46]

For Geldsetzer, the question is whether a logical connector may properly connect a position in the conceptual pyramid with itself.[47] He denies that it can, at least as a

---

[41] Geldsetzer, *Grundriss der Pyramidalen Logik,* p. 146.

[42] See text accompanying Fn. 11–12.

[43] 1006 a 31–35, 1006 b 14–17.

[44] Graham Priest, *Logic: A Very Short Introduction,* Oxford University Press 2000, p. 64.

[45] *An Essay Concerning Human Understanding,* Dover Publications Inc., New York, N. Y. 1959, 2 vols., Book IV, Chapter 8, 2nd. Voll. II: "Of Trifling Propositions", pp. 292–302.

[46] "Identity and Necessity", pp. 478–513, in *Philosophy as It Is,* ed. by T. Honderich and M. Burnyeat, Penguin Books 1979, p. 478.

[47] *Grundriss der Pyramidalen Logik,* p. 61.

logical matter; its use in mathematics, as we shall see, is a separate question. His argument for this position is that there is no sensory basis for such a self-reflexive logical connector; one doesn't see how an a concept or an object can be connected with itself. Identity is properly restricted to intensions, and the relation between positions in the pyramid can never be that of identity, but at most equivalence – and that, as we have seen, implies both identity and difference.

The present book contains several specific criticisms of this reflexive principle of identity, which I briefly summarize here. First, Geldsetzer brings out its contradictory character by reference to Leibniz's law of the indiscernibility of identicals. According to that principle, Leibniz defined identity ("x = x") as the indiscernibility of the two "x's" on each side of the equals sign. At the same time, however, he distinguished the *left* x from the *right* x. The contradictory character of the expression lies in this simultaneous determination of identity as both identity and non-identity, 7.1.5.

Second, Geldsetzer argues that the reflexive expression x = x cannot be considered a proper equivalence or equation. Using Frege's terminology, one can understand an equation as relating different signs which have different "senses" but the same "reference". On that model, the x to the left of the equals sign should have a different sense than that to the right of the equals sign. But in the tautology both signs have the same sense. Further, on the Fregean model the common "reference" of the two signs should be distinct from their respective senses. But in the tautology this distinction cannot be drawn; sense and reference are the same. Again, the attempt to construct a self-connection fails, 3.1.3.

Geldsetzer also suggests that attempts at reflexive self-connection run counter to the intrinsic properties of logical symbols as signs, leading to further contradictions. This can be seen on either of two readings of the expression x = x. First, "x" can be understood purely as a formal sign, signifying "nothing". On this assumption, "x" will mean "something" only after it has been applied to a number or a concept ("3 = 3" or as one usual says: "an apple is an apple"). But as a sign, "x" must denote something or it will fail to be a sign; so that in the reflexive expression of identity x is simultaneously understood as both a sign (because supposed to denote something) and not a sign (insofar as it is understood to denote "nothing"). Alternatively, "x" might be understood as referring simply to itself, as a sensorily perceptible sign. But in this case again we are compelled to simultaneously understand "x" as both a sign and not a sign; for signs intrinsically refer to something other than themselves, and yet here "x" is supposed to be a sign and nevertheless to refer solely to itself, 3.7.4.

10 Contradictions and Geldsetzer's Empiricist Program

In carrying through his empiricist program and pressing his attack on the supposed independence of logical forms, Geldsetzer arrives at an understanding of the nature and role of contradictions which is in many respects new. One of the innovations which allows him to do so is his treatment of concepts not as irreducible atomic units but as structured relations between intensions and extensions. This enables him to define conceptual contradictions as the fusing together of sensory perceptions which are mutually exclusive – contradictory – in the sense that they cannot be perceived at the same time in the same sensory context. His favorite example is "mortality". Logic textbooks frequently point to the necessity of concluding from "All men are mortal" that "Socrates is mortal". And this seems perfectly correct; we experience human beings – and animals generally – as both dead and alive, and so we understand "mortal" as meaning "alive (now or in the past) but destined to die". Geldsetzer, however, calls attention to the specific nature of the relations between the two sensory perceptions "dead" and "alive" and to the irregular structure of the concept "mortality" in which they are combined. What is decisive, from his empiricist point of view, is that while we surely experience human beings as both "dead" and "alive", these sensory perceptions do not and cannot occur in the same sensory context, that is, simultaneously in one and the same person. It is one or the other, and that is the root of the contradiction, 1.14.

Geldsetzer regards such contradictions as inherently ambiguous, both with respect to their meaning and their extensions, 1.14.5. Since they offer a choice of mutually exclusive perceptions, it will make a difference, for the truth or falsity of what we assert, which of these we choose to invoke. We usually have no difficulty in resolving this ambiguity through conventional agreement on one or the other of the meanings included in the contradictory concept. When we speak of Socrates as mortal, for example, we understand that he is physically long since dead, but lives on in our memory. But as a logical matter, either of the meanings of the contradictory concept is in principle available, and the result is that propositions including them are in principle indeterminate as to truth or falsity. The contradiction itself simultaneously allows both true and false propositions. In its most spectacular fashion, this occurs in paradoxes, which we will discuss below. Geldsetzer's thesis is that contradiction in this sense provides the best account of some of the most fundamental and puzzling concepts in logic, philosophy, and science, such as "possibility", "probability", "dispositional" concepts involving change and movement, as well as fundamental mathematical concepts such as " number" and "set".

This understanding of contradiction is at the core of Geldsetzer's empiricist challenge to rationalism. The rationalist position has been that logical consistency – the absence of contradiction – yields knowledge which need not await observation to claim certainty. We cannot think, according to the rationalist, a contradictory concept such as "round square". And because we cannot, we can know prior to the test of experience that we will encounter no such thing. Geldsetzer replies that thought cannot be separated from perception; we cannot think a round square in a regular concept, which comprehends only similars, but we can indeed think contradictions "in terminis"; however not as regular concepts, but as irregular ones which combine mutually exclusive perceptions of different round things or square things.

The rationalist also classically denies that sensory perceptions can account for the conceptual structures we bring to bear in experience. The argument is that indispensable concepts are not reducible to sensory experience. In reply, Geldsetzer suggests that contradictory concepts are a fundamental source of intellectual creativity; they enable us to transcend the limitations of particular sensory perceptions in both understanding and action without appealing to non-sensory entities, 1.14.7. His favorite example is the notion that the ancients had of a winged horse, the Pegasus. In actual experience, there are birds (with wings) and there are horses (without them), and the Greeks could not have experienced the fusion of the two. In constructing a concept which combined them, they used only sensory images obtained from experience, but nevertheless created a model which went beyond it. And that model could serve as a guidepost for research (mustering collected and new experiences to determine in what respects, if any, they correspond to the model) and invention (attempting to breed an animal which actually corresponded to the Pegasus). The contradictory concept, in short, can serve as a framework for our experience and as an orientation for extending it, and that is precisely what rationalism has always contended that sensory perceptions alone are unable to do.

11 The Dominant Logical Conception of Contradictions

A brief summary of long-prevailing logical doctrines concerning contradiction will be useful to put Geldsetzer's theories in perspective. The most important of these is that a contradiction is by its very nature, that is, by virtue of its form, false. Indeed, they are often said to be more than false, to be meaningless, or "absurd". Contradictions lack meaning, it is often felt, because the conflict they contain results in nullity: "[A] man who contradicts himself may have succeeded in exercising his vocal chords. But from the point of view of imparting information, of communicating facts (or falsehoods) it is as if he had never opened his mouth. He utters words, but does not say anything … Contradicting oneself is like writing something down and then erasing it, or putting a line through it. A contradiction cancels itself and leaves nothing".[48] As we shall see, the notion that contradictions are meaningless rests, from Geldsetzer's perspective, on confusion between intensions on the one hand, and concepts or objects on the other. It is the former which he sees as the irreducible units of meaning and identity, not the latter.

But in the logical tradition, the contradiction has largely been the formal hallmark of falsity. And in this capacity it serves as the basis of all "reductio ad absurdum" arguments. Their strategy is to attack a proposition not by attacking its premises, or challenging its truth or falsity directly, but by showing that, the proposition in question once accepted, contradictory consequences follow by valid forms of inference. There are distinguished precedents for such arguments in the history of philosophy. According to Plato, it was the form taken both by the attacks of Parmenides' opponents on his denial of plurality, and by Zeno's famous defenses

[48] P. F. Strawson, *Introduction to Logical Theory,* 1952, pp. 3–4.

of Parmenides' doctrine. The explanation Zeno gives to Socrates of his paradoxes is that they are meant to pay Parmenides' opponents back with interest, for they had attempted to show that Parmenides' axiom that all that is, is an undivided unity, led to ridiculous consequences; Zeno claimed to have shown that consequences still more ridiculous follow from the contrary assumption that that which is, is many.[49] In this sense, the contradiction plays a positive and indeed indispensable role in formal logical reasoning, as usually understood, because there is no other way of formally showing falsehood.[50]

Geldsetzer's thesis, however, is that all propositional contradictions are not wholly false, but mixtures – and extremely useful mixtures – of truth and falsity, and that contradictions "in terms" have (because they are concepts, not propositions) no truth values at all, but may have the important property of rendering the propositions in which they are employed simultaneously true and false. And in that connection he notes that Aristotle's writings on the subject – although usually cited as the foundational authority for the falseness of contradictions – are in fact consistent with such an understanding. Aristotle's fundamental formulation in the *Metaphysics* of what has become known as the law of non-contradiction does not exclude that opposing propositions may be true and false: "It is impossible for the same thing at the same time to belong and not to belong to the same thing and in the same respect."[51] In *De Interpretatione* Aristotle explains: "An affirmation is a statement affirming something of something, a negation is a statement denying something of something … It is clear that for every affirmation there is an opposite negation, and for every negation there is an opposite affirmation … Let us call an affirmation and a negation which are opposite a contradiction."[52] Aristotle also distinguishes genuine contradictories ("dogs are animals/dogs are not animals") from contraries ("dogs are black/dogs are white/dogs may be other colors"). "Contradictory opposites divide the true and false between them… Contrary opposites … are mutually inconsistent but not necessarily exhaustive; they may be simultaneously false, though not simultaneously true."[53] Thus, for Aristotle, what distinguishes the two forms of opposition is what came to be known as the law of the excluded middle: Each of the paired contradictory propositions must be either true or false, and there is no third truth value possible apart from these two (Latin: "tertium non datur", literally, "a third is not given").

As Geldsetzer points out, Aristotle nowhere says that propositional contradictions are false or lack meaning. He says only that the two propositions which constitute the contradictory pair cannot both be true, or both be false, at the same time and in the

[49] Plato, *Parmenides*, 127 d–e.

[50] Geldsetzer, *Logik*, (Scientia Verlag) Aalen 1987, p. 79.

[51] *Metaphysics*, 1005 b 18–20, transl. by Richard Hope, University of Michigan Press, 1960.

[52] *De Interpretatione,* 17 a 25–35, cited in Laurence R. Horn, "Contradiction", *Stanford Encyclopedia of Philosophy*, http://plato.stanford.edu/entries/contradiction/.

[53] Horn, "Contradiction", *Stanford Encyclopedia of Philosophy,* http://plato.stanford.edu/entries/contradiction/.

same respect, at least in the case of presently existing things.[54] This suggests, though Aristotle does not say so explicitly, that one of the contradictory pair of propositions must be true and the other false. And this is the position Geldsetzer adopts.

From the rationalist perspective, the falsehood of contradictions is often thought to have ontological implications. That which is contradictory is often said to have no basis in reality, no objective reference; contradictory concepts, such as "round square" are thus thought to be incapable of referring to anything. Contradictory concepts were one of the categories of "non-existent" entities whose status was much debated at the beginning of the twentieth century. The challenge such non-existent and self-contradictory entities posed (How can we meaningfully think entities which do not exist?) seemed (and still seems to many) to have been adequately dealt with by Russell's famous theories of descriptions and of types. In the theory of descriptions, non-existent entities were said to be disguised descriptions generated by the use of quantifiers which themselves derived their meaning from their function within propositions. Thus, "there is no golden mountain" became "there is no x such that x is mountainous and x is golden". In the theory of types, discussed briefly below, self-contradictory entities were allegedly dissolved by recourse to higher levels or types of entities. By means of these theories, Russell had succeeded, many thought, in domesticating non-existent entities and many bothersome contradictions.

On the one hand, then, the contradiction plays a positive role in logic as the formal hallmark of falsity. On the other, however, it appears as a threat to logical validity. Logically valid arguments are usually thought to be so because they preserve truth. But it has also often been held that any proposition, either true or false, follows from any false proposition. The Latin phrase is "ex falso sequitur quodlibet". But this can only be the case when the "false" here means "contradiction". Then "ex contradictione sequitur quodlibet"; that is, from a contradictory proposition, a true and a false proposition may be validly derived. From Geldsetzer's standpoint, this should be understood in light of the fact that contradictory propositions contain both truth and falsity. Thus, in the modern propositional calculus, where p and q are any propositions, a contradiction is symbolized as $p \& {\sim}p$; from there one may advance to p, from there to $p \vee q$, from $p \& {\sim}p$ again to ${\sim}p$, and, combining ${\sim}p$ and $p \vee q$, to q. But this limitation has not prevented many logicians from inferring "quodlibet" (what you will) from propositions which are simply false, rather than from contradictory propositions which are both true and false.

Moreover, as noted above, in the propositional calculus "logical" consequences are divorced from any thematic meaning of the propositions involved. But can it really be valid to argue that a false statement such as "Paris is not the capital of France" and a true statement as "Paris is the capital of France" follow from a

[54] Pierre Abélard tried to show that Christian dogmata are contradictions, and non-dogmatic assertions are not. See Geldsetzer, "'Sic et non' sive 'Sic aut non': La méthode des questions chez Abélard et la stratégie de la recherche", in *Pierre Abélard: Colloque International de Nantes,* ed. by J. Jolivet and H. Habrias, Rennes 2003, pp. 407–15.

contradiction, such as "Tigers are mammals and not mammals"? Such problems have given rise to an enormous amount of logical discussion, among the proponents of paraconsistent and relevance logics, among others.

12 Geldsetzer's Theory of Contradictions

With this briefest of glances at common logical conceptions of the nature and role of contradictions, let us turn back to how Geldsetzer develops his own. We noted above that Geldsetzer regards a contradictory concept as one which incorporates in a certain way perceptions which we cannot experience in the same sensory context, such as of human beings as "living" and as "dead". Several points are important here. First, the contradictory concept as "irregular" must be distinguished from "regularly" formed concepts, which embrace only perceptions which are similar, in the sense that various impressions of what is "living" are similar but that impressions of the "living" and of the "dead" are not. Geldsetzer is far from maintaining that all, or even most, concepts are contradictory. Regular concepts, too, subsume more particular concepts, which are constituted by specific differences, under themselves. Under the concept of tree, for example, there are concepts of specific kinds of trees, for example, conifers and deciduous. The difference between the regular and irregular concept is that the latter, unlike the former, contains the mutually exclusive specific differences within its own meaning, as the regular concept "tree" does not. One can think of the characteristics of a tree as such, without necessarily thinking of the differences between needles and leaves, 1.14.8.

Second, to understand what Geldsetzer means by "contradiction", one should recall both his empiricism and his theory of negation. "Not-A" means that which is different from or other than "A" in the sense that it cannot be perceived at the same time and in the same respect as "A", but it is always itself something perceived and as such positive. This negation only obtains between coordinate concepts. In this sense, negation may be either indefinite or definite, depending on whether it is located between coordinate concepts which are mutually exclusive or multiple. Negation, therefore, may be vague or determinate, but in either case it is a reference to something positive. "Not-red", without further specification, that is, as an indeterminate negation, may mean "green" or "some other color" (or sometimes "something other than color"). Determinate negation can obtain only between dihaeretic (mutually exclusive) concepts, as in A vs. non-A. A contradictory concept, then, is a fusion of characteristics which are contradictory in just this sense of mutually exclusive perceptions.

This should be distinguished from another, non-empirical notion of contradiction. On this latter view, it might be the case that one does not find anything positive which can be understood as the contradictory opposite of a given concept. Then "the contradictory of 'X' would be 'non-X', that which is intrinsically opposite to X; then no Absolute, however generously encompassing, could enfold both 'X' and 'non-X' into its unifying embraces."[55] It is difficult to say precisely in what such an "intrinsic opposition" among concepts could consist, but such a view seems only compatible with the rationalist and realist conceptions which Geldsetzer combats.

Third, the contradictory concept has as its extension the extensions of its constituent concepts. It does not, as is often thought, refer to nothing; it refers simultaneously and in the same respect to each of the mutually exclusive concepts which constitute it 1.14.8. "Pegasus", for example, has as its extension birds to the extent that they have wings, as well as horses, as both of these are perceived, recollected in memory, and combined in imagination to form the contradictory concept. As the example shows, it may be that there is no actually existing Pegasus, or none yet. But the concept has nevertheless its extensions; what it lacks is a unique extension of its own distinguishable from the separate extensions of its components.

Fourth, as a result of its structure, the contradictory concept, when employed in a proposition, is as a logical matter ambiguous, since either of its mutually exclusive component meanings may be invoked, 4.6., 4.6.1. When we speak of Socrates as mortal, we understand that he is now dead but was, when alive, not dead but fated to die; when we speak of ourselves as mortal, we realize that we are alive, but that our deaths – which we are not capable of experiencing at the same time as our lives – must come. It is this oscillation between the mutually exclusive experiences built into the structure of the contradictory concept which results in both true and false propositions.

Fifth, Geldsetzer's view is that it is intensions, not concepts, which can be identical. This requires that we carefully differentiate two concepts which have sometimes been equated – identity and contradiction. Leibniz, for example, wrote that "[n]othing should be taken as first principles but experiences and the axiom of identity or (what is the same thing) contradiction, which is primitive, since otherwise there would be no difference between truth and falsehood, and all investigation would cease at once."[56] The result has been that concepts and objects have been conceived as reflexively self-identical, as we saw above. It seemed true and necessary to say that an apple is an apple; for if it were not, how could we speak and think meaningfully about apples?

Geldsetzer's reply to this is that the unit of meaningfulness and identity is not the concept but the intension. In the case of the regular concept, the intensions contained within a concept will be consistent, in the sense that they all arise from what can be perceived at the same time and in the same context – for example, an apple. But in the case of some important concepts, as we will see directly, the intensions which the concept contains are incompatible. Each of them, nevertheless, is separately perceptible and understandable. That is why we are able to grasp and understand that a contradiction exists, and why it is wrong to say that contradictions

[55] John Passmore, *A Hundred Years of Philosophy,* 2nd ed. 1966, p. 69.

[56] Leibniz, *New Essays Concerning Human Understanding*, ed. and transl. by A. G. Langley, 2nd ed. 1916, pp. 13–14, cited in Horn, "Contradiction", *Stanford Encyclopedia of Philosophy.*

are "absurd" or "meaningless." That is also why, as we attend to one intension or the other, the contradictory concept may form the basis of propositions which are alternately true and false, so that the contradiction is properly thought of as a mixture of truth and falsity.

It might be objected that, so long as we can distinguish their components, such concepts are not contradictory at all. "[W]e might apply to the same thing, in the same breath, two predicates, which would ordinarily be regarded as incompatible, without contradicting ourselves. If we do this, we invite the question 'What do you mean?': and if we can explain what we mean, or show the point of saying what we say, then we have not contradicted ourselves. But if there is no way of doing this, we are inconsistent."[57] Geldsetzer's point is that in the case of some genuine contradictions we cannot explain what we mean by appealing to a unitary sensory perception; to explain the contradictory concept we must necessarily invoke opposed sensations which cannot sensibly or empirically be combined at the same time and in the same respect. Strawson's example illustrates this: "Thus we might say, in answer to a question, 'He is both over six foot tall and under six foot tall', and then explain that he has a disease which makes him stoop, but that if he were cured and were able to stand upright, he would top the six-foot mark." The contradiction is resolved, Strawson suggests, by appeal to the concept of possibility, but that is itself a concept which Geldsetzer shows to be contradictory, as we shall see.

Geldsetzer's view diverges radically from another current doctrine concerning contradictions, dialetheism, which "is simply the view that some contradictions are true."[58] Dialetheism, that is to say, holds that "there are some sentences (statements, propositions, or whatever one holds truth-bearers to be), a, such that both a and ~a are true." With this, Geldsetzer does not agree. He stresses, on the contrary, the mutually inconsistent character of the propositions which flow from contradictory concepts. It can never be the case that some contradictions are simply true; contradictions are themselves mixtures of truth and falsity. Let us turn to some examples.

13 Examples of Contradictory Concepts

What such a theory of contradictions can accomplish Geldsetzer shows through his proposed reconstructions of key concepts in philosophy, logic, theory of science, and mathematics. Some examples may illustrate that briefly here. "Possibility", among the most prominent of concepts in both logic and science, is usually considered as a regular concept. Its subordinate species are both the actual world, which is understood as one among many possible worlds, and various non-existing but nevertheless "possible worlds", 1.14.8. On this view, "possibility" is the common generic concept of that which is – being – and that which is not – nothing. Such a

[57] Strawson, *Introduction to Logical Theory*, p. 7.

[58] Graham Priest, *In Contradiction*, 2nd ed. 2006, p. 299.

regular concept presupposes that its common generic characteristics can be thought in abstraction from – disregarding – the specific characteristics of "being" and "nothing", 1.14.3. But "possibility" means nothing other than "being and nothing"; it encompasses nothing other than precisely these mutually exclusive characteristics. Everything possible both is and is not. Nothing else remains as the positive content of the concept of "possibility". The possible arises when we perceive something – which accordingly is – and combine it in contradictory fashion with what is not; in short, we construct possible worlds on the basis of perceptions. "Possible worlds are not worlds in being, but to the extent they have been conceived and constructed (particularly by logicians) they are also not nothing", 1.14.3; see Figure 5 at 1.14.8.

Dispositional concepts, which express capacity and propensity to change, are a second example. Carnap, and many since, have believed that they resist reduction to an empirical basis. They are employed with particular frequency in psychology and the natural sciences, and include such concepts as capability, capacity, force, disposition and power. The difficulty for empiricists has been the inability to point to a perception or set of perceptions which seems equivalent to the capacity for change in a particular direction which such concepts express. Geldsetzer proposes to treat them as contradictory – or, more precisely, because they arise from the fusion of two concepts within a row of mutually exclusive coordinate concepts – contrary concepts, 1.14.9. and Figure 6. Take the problem, for example, of constructing, on an empirical basis, a concept which expresses the capacity of a substance to pass from a solid to a liquid state. We say that it is liquefiable under certain conditions, but what is the perceptual basis for this assertion of liquifiability? We observe the substance in a solid state, and then in a liquid one; but the capacity to change from one state to the other seems to have no perceptual correlate. Geldsetzer's suggestion is that the extension of such a concept consists of the extensions of the two concepts which it unites – solid and liquid. These exclude each other; they cannot be perceived at the same time and in the same sensory context. That is why uniting them results in an irregular concept. But again, as in the case of the contradictory concept in the narrow sense, a dispositional concept cannot be thought if the mutually exclusive characteristics which make it up are disregarded; being liquefiable is not something separable from being solid and then being liquid; that is all that it is.

14 Probability as a Contradictory Concept

Propositions expressing probabilities are nearly dominant in modern science; Geldsetzer is critical of this dominance, 4.7. He regards as mistaken claims that properly drawn probable inferences may be "closer to truth than falsity". If that were so "there would be no games of chance, gambling, lotteries or stock market speculation", 8.9.2. He regards them instead as mixtures of real or supposed knowledge and ignorance, 4.7.3. It has been said that in a probable inference, certain probabilities are transferred from the premise to the conclusion; Geldsetzer denies this, 5.9.1.

He argues instead that such inferences express a mixture of knowledge concerning the structure or interdependence of events with "absolute ignorance concerning the forthcoming of a singular conjectured or predicted event", 4.7.3. "Methods of quantifying probabilities merely trace the periphery of knowledge around a point where nothing at all can be known, namely the occurrence or non-occurrence of a single event or the presence or absence of a single instance", 4.8.10.

The conceptual structure of propositions concerning probability is contradictory. Possible outcomes are mutually exclusive, same-level species concepts of a common generic concept of a single event. Logical probability, as Geldsetzer terms it, is always symmetric between two concepts fused in the concept of probability. That is visibly demonstrated by 1 : 1 or 50 : 50 probabilities. For example, the common generic concept of a coin toss has subordinate, mutually exclusive species concepts of a coin toss resulting in heads and a coin toss resulting in tails, 4.8.3. These two species concepts may be fused to yield the contradictory concept of a coin toss with probability of one-half for heads and one-half for tails. This is contradictory in the sense described above; that is, the concept of the "probable" outcomes has no intensional content apart from that of its contradictory components. And its structure combines assumed knowledge as to the structure of the outcome (the result will be either heads or tails, one does not anticipate that the coin will remain standing on its rim) with ignorance as to the outcome. This ignorance is masked by the fact that both of the mutually exclusive outcomes are predicted; both will come up, but each of them with probability of 50%, 4.7.5.

As a logical matter, therefore, one cannot err in making a statement of probability, for both (or multiple) outcomes are asserted. The situation is not fundamentally different when probabilities are mathematically formalized and extended to multiple result concepts, 4.8.2. This usually involves the calculation of projected relative frequencies, that is, the relationship of one predicted event to all possible events of the same class, so that a roll of a die yields a probability for any number from one to six or 1 : 6. Some mathematicians have also – correctly – remarked that an adequate notation of mathematical probability should not set the projected event in proportion to all possible events, but rather only to its counter-instances. Then the probability of a particular event in the case of the roll of a die would not be 1 : 6, but 1 : 5. What this shows, however, is only that such quantification of probabilities doesn't tell us anything useful for purpose of making predictions, let alone for estimating the precision of such predictions.

The inference usually drawn from such mathematical calculations is that they quantify "degrees of probability" which are transmitted from premises to conclusions. But this, Geldsetzer says, "fails to take into account and cannot explain that and why in each such calculated probability the particular event either occurs or doesn't", 4.8.2. And this in turn means that logical probability obtains even in cases of mathematical probability.

None of this means that we do not frequently have good reasons for expecting one thing to happen rather than another. What is mistaken is the impression that a proposition to which a higher degree of probability is attributed approaches more closely to a true prediction, 4.8.1. What does vary is the quantum of knowledge which may be presupposed by and contained in the proposition, based on our relevant past experience. But that quantum of knowledge, no matter how great, does not result in a greater or lesser proximity to what one doesn't know. "That is why even that which is most 'highly probable' just as that which is least probable still either occurs (turns out to be true) or doesn't", 4.8.1.

Probable inferences are disingenuous to the extent that they appear to immunize scientists against error. They seem to do so because they assert both the occurrence and the non-occurrence of the event of interest. And that means that they are contradictions, and not genuine assertions at all. When formulating predictions Geldsetzer recommends, in the place of the indicative mode proper to assertion which now predominates ("It is probable that this or that will occur") that one substitute the subjunctive, the mode of conjecture ("This or that could occur"). In science as in daily life we may have good reasons for a causal or correlative assertion of the form: "When a certain state of affairs obtains, another specified state of affairs will follow". But such assertions cannot avoid the risk of error. That is what distinguishes them from the contradictions inherent in probability. These latter have the result, Geldsetzer satirically remarks, that no scientist ever makes a mistake.[59]

15 Contradictions, Propositions, and Theories, Especially Mathematical

We began by outlining the structure of contradictory concepts and with our discussion of probability have already begun speaking of contradictory propositions. Let us turn back for a moment to consider how contradictory concepts function within propositions and then look at the role contradictions play within scientific and especially mathematical theories.

A contradictory proposition links a contradictory or irregular concept together with regular, non-contradictory concepts by means of a proposition-forming connector in such a way that the linkage may be read as both true and false, 4.6. When we say that a state of affairs is possible, for example, we mean that it both is and is not; and so it is both true that the state of affairs is, and true that it is not, and false that it is, and false that it is not. These contradictions are usually not recognized as such, 4.6.1. They are masked both by conventional acceptance of one of the branches of the contradiction and by scientists' willingness to switch from one to the other when the context so requires, 4.6.1.

A contradictory proposition is therefore irreducibly both true and false, because of the mutually exclusive perceptions which it juxtaposes. This view collides head on with the generally accepted view of logicians, discussed above, that contradictions (AB is A and not-A) are false by their very nature – and that means by virtue of their form. And it is equally at odds with the dominant logical notion that an

[59] Geldsetzer, *Grundriss der Pyramidalen Logik*, p. 90.

alternative, one branch of which is true and the other false (AB is A or not-A), is true by virtue of its form. In both instances Geldsetzer rejects the notion that there is a formal, logical meta-level which warrants abstracting from the contents of the contradiction or alternative.

The attempt to carry through this separation of form cannot succeed. For we can make no sense of the notion of a contradiction without recognizing that it is nothing other than a mixture of truth and falsity. When we say, for example, that the weather will change and that it will stay the same, there is no contradiction unless we recognize not only that what is asserted is mutually exclusive, but also that one of the alternatives will prove to be true, and the other false. Without the recognition of truth and falsity, there is no contradiction. The claim that the form or structure of the proposition can be conceived as a meta-level of meaning which makes the contradictory proposition false independently of these contents does not hold. And similarly the alternative is not true by virtue of its form alone but by virtue of the fact that one of its branches is true, and the other false (although, of course, we may not know which is which). This is the structure of many tautologies, which are thought to assert nothing, but do in fact assert both what is true and what is false. (For example, the weather will change, or it will stay the same.) Absent such a mixture of truth and falsity, there is no genuine alternative or contradiction at all.[60]

When contradictory concepts are used in scientific theories, their presence "shows itself in the construction of propositions which contradict other propositions within the same theory", 6.3.6. The point applies generally to theories in logic, philosophy, and the sciences, and Geldsetzer provides examples of both intentional and unintentional contradictions within theories. Here I want to mention only one aspect of particular importance: the role of the contradiction in understanding mathematical theory and mathematical logic and in determining their relationship to classical logical principles. Geldsetzer can fairly be termed a logicist with regard to mathematics, in the sense that he holds mathematics to be reducible to logical principles. But he defines the relationship between the disciplines in a new and surprising way. What properly distinguishes mathematics and mathematical logic from classical logic, he tells us, is mathematics' dialectical character – that is, the methodological use which it makes of contradictory concepts. Certain of the basic principles of mathematics are indeterminate as to their truth and falsity because "they consist of contradictory basic concepts and propositions formed with them from which both the truth and falsity of mathematical theories can be deduced", 7.2.

In a previous network publication,[61] Geldsetzer has given a fuller account than is possible in the present book of how mathematics (and mathematical logic) came to be what he views them as: the (unacknowledged) systematic elaboration of

---

[60] L. Geldsetzer,*Grundriss der Pyramidalen Logik*, p. 86.

[61] L. Geldsetzer, *Grundriss der Pyramidalen Logik*, (Internet HHU Duesseldorf, 2000) in summary overview at pp. 17–19, and more extensively at pp. 125–48 (the chapter is entitled "Euclid's Foundation of Mathematics on the Basis of Platonic Logic").

contradictory concepts. He begins with an analysis of Euclid's famous *Elements.* It was for centuries, continuing into modern times, the fundamental mathematical textbook, which not only supplied basic geometrical and mathematical concepts and principles, but exemplified and inculcated what became the mathematical method of thought. Those concepts and that method, Geldsetzer argues, can best be understood in light of Euclid's pervasive but unacknowledged adoption of Plato's doctrine of ideas. That doctrine, in Geldsetzer's interpretation, is that on the one hand, signs, symbols, sensory percepts generally, point beyond themselves to something which, in sensory terms, they are not – something which can only be grasped in thought; but that, on the other hand, we can only think – put into conceptual form – that which we have already perceived.

Euclid's exposition begins with perceptible models, and suggests that these are sufficient for the understanding of the geometrical and arithmetical concepts which he proceeds to construct. What is Platonic and dialectical in Euclid's method, Geldsetzer points out, begins with the fact that these perceptual models themselves do not and cannot fully capture the content of Euclidian concepts; a Euclidian "point" is something different and other than a perceptual point, and similarly with a Euclidian "line". So far, this seems more consistent with rationalism than with empiricism, for Platonically inclined mathematicians have always believed that mathematical entities are more than any sensory model which may be used to elucidate them.

But Geldsetzer provides a detailed demonstration, through an analysis of many of Euclid's concepts, that the "something other" and "something more" which inheres in such concepts can best be understood in empirical terms, as contradictory conceptual constructions. Euclid defines a "point" for example, as "something which has no parts". Everyone is familiar, as a matter of sensory perception, with points in the sense of very small patches of color, or as intersections of two lines. But these can and often do have plainly perceptible parts, and therefore are not a sufficient empirical basis for the Euclidian concept of "point". That concept can be best understood as the fusion of the perception of experienced points, which have parts, with a perception of something which does indeed have no parts, such as an atom or individual, defined by Democritus as a "partless entity". Euclid fused these inconsistent perceptions – of a perceptible point which has parts, and of an entity which doesn't – in his concept of "point". That concept cannot, as mathematicians have understood it, be directly sensorily demonstrated but only thought – but it can and must be constructed out of different sensory contexts into an imaginary unity.

Euclid constructed arithmetical concepts in the same fashion. The concept of "number" is contradictory in the same sense; it is the fusion of two sensorily incompatible elements of perception: individualization (for each number denotes the unity of an element expressed in "a" or "one of") and universalization (since each number also and simultaneously expresses the totality implied in "all"). "In experience, we can fix our attention sequentially either on 'an' object or on 'all' objects within a given range of perception, but only in recollection or in fantasy can we simultaneously represent to ourselves 'one and all'. That is, however, precisely what is required in a representation of number, namely that one thinks 'all (the totality) as one' or, in reverse, 'one as (a totality) of all' ", 1.15.2. No regular concept of number is possible, for there is no characteristic common to both "one" and "all" which could separately supply the content for such a concept, in abstraction from unity (or elementarity) and totality (of elements). Use of the concept of number therefore necessarily generates both true and false propositions. What is true of a number considered as "one" will be false with respect to the same entity considered as a "many", or a "totality of all" of its parts. And vice versa. We do not generally notice this in arithmetical operations because mathematicians customarily take into account only one part of these contradictory characterizations.

The genius of such contradictory concepts, and of Euclid's presentation of them, is that they offer, to the beginning student, familiar perceptual models from which to begin the process of thought, just as Plato himself offered parables and myths to ease the ascent to the realm of ideas. At the same time, however, each contains opposing and inconsistent perceptions, so that the concepts themselves can only be grasped through the synthetic imagination. For those capable of this, however, for the mathematical initiates, the reward is that either of the inconsistent perceptions which each concept contains is available for use, as the occasion demands. And it is this, Geldsetzer suggests, rather than any mystically tinged rationalist conception of "pure" intellectual thought, which accounts for the way in which mathematical thought appears to transcend sensory perception.

Nevertheless, the dialectical-contradictory structure of mathematical thought has seldom been acknowledged, and has largely led a subterranean existence in the history of logic and philosophy. It was Leibniz, Geldsetzer argues, who revived and developed this Platonic and dialectical method of contradictions while combining it with the newly developed methods and basic concepts of modern mathematics and natural science. The modern mathematician, natural scientist, and mathematical logician follow the same program, which consists in combining inconsistent perceptions into key concepts and exploiting one or the other as the occasion demands, usually without being aware of it. For their purposes, as Geldsetzer puts it, "the continuum is discrete and the discrete continuous as needed; the infinitely small vanishes into nothing, but is at the same time something; the infinitely large is the all, but at the same time something and therefore again at the same time a smaller "cardinality" among still greater "cardinalities" (G. Cantor); the set is a multiplicity of elements (as understood in the "logical" concept of the set) and yet remains a set, even if it contains only one member or only itself as a member, or no members at all (the empty set); mathematical concepts, structures, and entities are fundamentally imperceptible, but at the same time also perceptible through 'instantiations', interpretations, applications, and models." Consequently, Geldsetzer argues, when we come to applications of mathematics in physics, it can't be surprising that these contradictory determinations are transferred to physical phenomena. Time becomes simultaneous/non-simultaneous and a location in space simultaneously determinate and relative (theory of relativity), the particle becomes simultaneously a wave (principle of duality), and quantum physics tells us that measurement of the movement of a wave/particle (which is defined via "velocity" as duration of time in proportion to extension of length) can also be taken at extensionless time-and-location "spots", with the rather mysterious result that one or the other component or both become indeterminate (Heisenberg's uncertainty principle in microphysics), etc.[62]

16 Set-theoretical and other Paradoxes

A similar analysis illuminates Russell's set-theoretical paradox, which was for Frege a crushing blow and which Russell intended his theory of types to overcome. It arises from the juxtaposition of the concepts "set which does not belong to itself" and "set which does belong to itself". As Frege put it, if the "set which does not belong to itself" is understood to come within its own extension, to belong to itself, we have a contradiction. For then the "set which does not belong to itself" does belong to itself. If on the other hand we suppose that the "set which does not belong to itself" does not fall within its own extension, does not belong to itself, we again have a contradiction, for then it does belong to itself. In either case, it does and does not belong to itself.[63]

Geldsetzer's thesis is that the paradox arises from the contradictory juxtaposition of two mutually exclusive concepts of "set". On the one hand there is the Euclidian and common-sense notion of set, according to which a set is a group of elements, but is consequently incapable of being "empty" or of "containing itself", just as everyday objects cannot. On the other there is a modern numerical concept of set understood as capable both of being empty (since zero is now understood as a number) and as capable of containing itself (and this understanding is a consequence of the self-reflexive understanding of identity discussed above). Again, there is no characteristic of "set" separate from these mutually exclusive characteristics, and therefore no regular mathematical concept of set. When one employs the contradictory concept of set in mathematics, then, one will generate both true and false propositions, for the same proposition will be true of the one kind of set and false of the other, and vice versa. A charitable interpretation of such mathematical propositions will focus on their truthful aspects, and disregard their falsity, so that the contradiction will not be readily apparent. But as a logical matter it is present, as Russell's set-theoretical paradox shows.[64]

Geldsetzer's theory of contradictions also offers an important new perspective on the nature of non-mathematical paradoxes such as Eubulides' "Liar". The paradox arises when, for example, someone says "I lie", so that the proposition appears to assert its own falsity. The dominant view among logicians and mathematicians is and has been that such paradoxes can be (and that some have been) solved. The solution usually consists in an attempted demonstration that the apparent

---

[62] L. Geldsetzer, *Grundriss der Pyramidalen Logik,* p. 18.

[63] W. and M. Kneale, *The Development of Logic,* Oxford University Press, 1968, p. 652–53.

[64] L. Geldsetzer, *Grundriss der Pyramidalen Logik,* p. 92.

contradiction is itself "false", and that consistency can be restored and insured by recourse to a meta-level of analysis, where one speaks of "types", "levels", or "functions", for example.

Geldsetzer rejects any notion of a meta-level and affirms, as we have seen, the irreducibility of the contradiction. Falsity is not something which can be appealed to in order to abolish contradictions; rather, contradictions contain truth and falsity as components, and contradictory propositions are true and false at the same time. That is indeed the peculiar characteristic of paradoxes. Geldsetzer quotes an ancient commentator on Aristotle, Alexander of Aphrodisias, who appears to have recognized this in relation to the Liar: "He who says 'I lie' lies and tells the truth at the same time."[65]

17 The Foundations of Logic

We began by noting the changes in notions of logical form and consequence which arise from Geldsetzer's refusal to separate logical forms from perceptual content or to accept the legitimacy of thought which claims to operate at formal "meta-levels" while dispensing with meanings. Let us now close by introducing the consequences of this view for the foundations of logic itself.

"The concepts identity, contradiction and the 'third' (neither identity nor contradiction) are considered the highest-level general concepts of logic", 7.1. In both classical and mathematical logic, these concepts "are employed as forms of notation of the so-called truth values ('true', 'false', and 'third and other truth values', as distinct from either true or false)", 7.1.3. The grounds on which identity, contradiction or the third are found to be present are understood to be, as discussed above, purely formal, as are the consequences which flow from them. Identical propositions are conceived as "formally", "logically", or "tautologously" true, and contradictory propositions as false in the same manner. Geldsetzer's view, as we have seen, is that this "formalist" conception of logical form is not serviceable for logical purposes.

Those purposes are the determination of truth, falsity and truth-falsity or probability. Truth and falsity are themselves the "highest-level regular generic concepts of logic and as such its conceptual axioms", 7.4. All logical propositions and inferences are located within their extensions, and they stand in a relation of negation to each other. The sole characteristic of truth is logical coherence; the sole characteristic of falsity is logical incoherence.

Since truth and falsity are the highest axioms auf logic, there is no way of abstracting from coherence and incoherence and arriving inductively at a still higher common generic concept. On the other hand, were truth and falsity the sole logical axioms the result would be a purely bi-valent logic. In order to take account of the logical fact of the 'third', that is, of contradiction in logical concepts and propositions,

[65] L. Geldsetzer, *Grundriss der Pyramidalen Logik,* p. 92.

there must be a proper axiom of contradiction. It consists in the "dialectical" fusion of truth and falsity to define "truth-falsity". "Truth-falsity or probability is a contradictory concept which arises from the fusion of truth and falsity … The characteristics of probability are coherence and incoherence", 7.5. This third axiom of logic is the matrix of all dialectical elements of both logic and of mathematics as they have developed and are currently constituted.

This brief introduction can give little impression of the extraordinary richness of the present book which ranges widely over the elements of logic and mathematics, drawing on the history of both disciplines and current controversies.

This publication is well suited to serve as an introduction, particularly for the English language reader, to Prof. Geldsetzer's work in the philosophy of logic and mathematics. It is a distillation, with some notable developments, of his earlier writings on these subjects, most importantly his *Logik* (Aalen 1987) and his *Grundriss der Pyramidalen Logik mit einer logischen Kritik der mathematischen Logik und Bibliographie der Logik* (Internet of the Institute for Philosophy of HHU Duesseldorf 2000), neither of which is yet available in English. And a valuable overview of the logical literature as a whole is provided by Prof. Geldsetzer's *Logic Bibliography up to 2008* (Internet of the Institute for Philosophy of HHU Duesseldorf 2008).

The current text is published here for the first time in English (although an earlier shorter German version under the title "Elementa logico-mathematica" appeared on Prof. Geldsetzer's homepage in 2006). In addition to the text proper Prof. Geldsetzer has prepared an extensive set of "Notes and Corollaries". These not only elucidate the propositions set forth in the text but through extensive citations of literature effectively situate Prof. Geldsetzer's views within the broader contexts of the history and contemporary study of logic and – to some extent – the philosophy of science. They also demonstrate – with impressive scope and detail – the extent to which Prof. Geldsetzer's theses rest on a deep knowledge of the logical and mathematical-logical tradition but also cast it in a new light and propose its substantial renewal.

The translation and this introduction have benefited greatly from intensive discussions with the author and from his review of drafts of both. Indeed, Prof. Geldsetzer prepared the initial draft of the "Notes and Corollaries" in English, so that the task of translation was reduced to one of rendering some passages into a more idiomatic English. It remains only to thank Prof. Geldsetzer for his friendship, kindness and generosity throughout and to note that the responsibility for any shortcomings in the translation and introduction is mine.

New York City Richard L. Schwartz

Chapter 0
Preliminaries

Logical form as the distinctive feature of the scientific enterprises – On logical formalisms and their relation to language – The pyramidal schema exemplified in Chinese ideographic characters – Mathematics as the dialectical outgrowth of Aristotelian "trivial" logic, developed historically in the "quadrivial sciences" (mathematical natural sciences) – Synopsis of themes 8

0. Logic is the methodology common to all the sciences (defined here to include the humanities). Its use within a science is an essential criterion of the scientific character of a science or discipline. A "science" without logic is not science at all.
   - 0.1. Within logic itself, what is most important is the logical formalism, for it is this which gives to thoughts their logical form.
     - 0.1.1. Most formalisms have been developed on the model of phonetically written languages, and it is these formalisms which have most frequently been used up to the present. Those logicians who for that reason believe logic to be an artificial language, with its own lexicon and grammar, consider logic itself to be an "ideal language."
     - 0.1.2. Some formalisms, in particular those used in so-called mathematical logic, have been developed on the model of arithmetical calculations. Those logicians who for that reason believe that one can "calculate" with concepts and in some cases with judgments and inferences – as Th. Hobbes, G. W. Leibniz, E. Bonnot de Condillac, G. Ploucquet, J. H. Lambert and G. Boole began to do – , regarded the essential logical formalisms as calculi (tools for calculating).
     - 0.1.3. Some formalisms have been developed as graphic representations on the model of geometrical diagrams. These include Porphyrian trees and their inversion to conceptual pyramids (which are used in this book), L. Euler's circles, J. Venn's diagrams, B. von Freytag-Löringhoff's pyramidal representations of some connectors, J. Sowa's as also B. Ganter's and B. Wille's graphs for the representation of

L. Geldsetzer and R.L. Schwartz, *Logical Thinking in the Pyramidal Schema of Concepts: The Logical and Mathematical Elements*, DOI 10.1007/978-94-007-5301-3_0,

"formal conceptual analyses". These are often used solely for didactic purposes, and are considered merely pictorial aids for the representation of logical relationships otherwise not accessible to sensory perception. But this customary appraisal of their value does not do justice to graphic formalisms. They are similar in many respects to pictorial characters (like the Chinese) which represent meanings by means of sensorily perceptible images. In Western cultures one is scarcely accustomed to use such characters. Perhaps the traffic signs now found throughout the world will gradually help us learn to do so. In this book we will place the graphic diagram of the conceptual pyramid at the center of the discussion and assign it new applications.

0.2. The formalism does not merely present an image of the logical, rather it is the essence of logic. Graphical formalisms are logic made visible – although not as picturesquely as A. J. Swinburne's "Picture Logic" of 1881 did. The study of Chinese ideographs – with attention to their "etymological" pictorial content (which they have essentially retained in addition to their phonetic representation) – is very helpful in becoming accustomed to this kind of representation. It is only as a remembered schema that graphical formalisms become an image of the essence of logic.

0.3. Formalisms and their parts – so-called logical symbols – never lack sense or meaning. "Signs" which didn't signify wouldn't be signs at all. The belief that formalisms or logical signs have no sense or meaning prior to interpretation and are first assigned sense and meaning through interpretation is therefore one of the most widespread misconceptions concerning logic. In fact, what we call interpretation or assignment of meanings *adds* a layer of specifically logical meaning to *prior* meanings – that is, the marks on the page must already have been construed as, for example, letters of the alphabet, numerical digits, or graphic designs – before their interpretation as logical symbols can proceed.

0.3.1. To logically formalize facts means to impress the structure of the formalism upon them. In language-analogue formalisms it is not yet genuine formalization, but only a kind of shorthand writing, in which words or concepts are simply replaced by single letters.

0.4. Logical thinking consists in this: That thoughts are ordered and connected with each other according to the structure and the meaning of the formalism.

0.5. Thinking presupposes that thoughts or ideas are evoked by means of language. That doesn't mean that logic is to be established on a "linguisticistic" or "psychologistic" foundation. It does mean that logic cannot be carried on at all without sensory perception, speech, memory and thought.

0.6. Logic adopts elements from language which it uses to construct its formalisms and to "read" or describe calculi and graphic formalisms. The selection of such elements makes logic a linguistic code within a developed language.

0.6.1. When they are used for logical purposes, letters (as signs designating certain sounds) and the logical connectors (junctors) retain their linguistic character as elements of language and presuppose the understanding of their meaning which occurs in language use. Similarly, the ordering on the page of formalisms which are analogous to language, the sequencing of logical symbols from left to right in lines, and the arrangement of the lines themselves under each other can only be understood by reference to their use in certain written languages.

0.6.2. The selection of further logical elements from the elements of language orients itself on the grammatical description of languages. Consequently, many grammatical concepts also occur in logic, as, for example, sense and reference, term, subject and predicate, sentence, quotation, sequence, theory of meaning or semantics, synonymy, homonymy, grammaticality, and syntax or syntacticality. Many specialized logics or formalisms distinguish themselves according to the linguistic and grammatical elements which they select for adoption. The grammatical subjunctive for the expression of conjectures (statements which are not assertions) is formalized in logic in the indicative, so that conjectures assume the form of statements which seem to be assertions. This is also the case in formalizations of subjunctive conditionals and so-called contrafactual assertions. Only the Stoics adopted the grammatical subjunctive for the premises of their inferences. But even these are regularly translated from the Greek in indicative form. The grammatical subjunctive is also the unacknowledged basis of statements expressing probabilities.

0.6.3. Quotations and the signs which designate them (quotation marks and mathematical parentheses) play a special role in logic and mathematics. Quotations in language have several functions, which are not always clearly distinguished from one another and employed separately. 1. Simple emphasis of the content of the quotation; 2. Suspension of any claim to truth or falsity made by an assertion; 3. The duplication or multiplication (effected in the latter case by quotation of quotations) of the meaning of the sentence (the distinction drawn here is that between the original sense of the content of the quotation and the meanings it may acquire through its use in arguments); 4. Suspension of the meaning of the content of the quotation itself; 5. The replacement of the content of the quotation through another linguistic or mathematical category (for example, the content of a sentence in a natural language can be replaced by the "concept of the sentence," or the mathematical content of a parenthetical expression of an addition through a "sum"). In medieval logic, some of these functions were distinguished as different "suppositions" of signs.

0.6.4. Only the first through the third functions play a role in classical logic. In modern mathematical logic generally and the propositional logic which has developed with it all of these functions are used.

This leads to essential differences between classical and so-called mathematical logic. The differences show themselves most clearly in the use of the logical particle "meta" in mathematical logic.

0.6.5. In classical logic, "meta" designated a logical flaw in an argument, expressed in the Greek phrase "metabasis eis allo genos" ("transition to another category"), that is, an unacknowledged change in the meaning of concepts employed in the argument. In modern mathematical logic, "meta" signals a dialectical trick: To make the transition from something which is given to something else, and at the same time to remain with what was initially given. In that sense, "meta" has the third through the fifth meanings attributed above to the use of quotation marks. Whoever speaks about a language in a "meta-language" takes that language itself to be an object, distinguishes it nonetheless at the same time as the "object language" from objects themselves, and even when speaking the meta-language remains – in order to be comprehensible – within the framework of the generally understood language of educated people.

0.6.6. It was in this sense that A. Tarski, in 1936, spoke of white snow as an object, while simultaneously asserting in the object language that snow is white (a statement which may be considered to be "analytically true"!). He then named this assertion – quoting it in the form of "the snow is white" – a "concept" of the "meta-language" and proved – persuasively, for most mathematical logicians – that logical truth consists in the possession of a concept in a "meta-meta-language" of the agreement between the concept of the quoted statement and the statement itself. With this thesis, Tarski believed, he had explicated the Aristotelian concept of truth as a (realistic) correspondence of being and knowledge about that very same being.

0.6.7. Mathematical logic obtains many logical elements by borrowing mathematical parts of the language of educated people and adding to these some genuinely mathematical elements. By the former means the number of synonyms employed in logical terminology has greatly increased. We speak, for example, of logical constants or functions (signs designating functions) instead of logical connectors, variables instead of signs for concepts or judgments, class or set instead of extension, logical sum instead of disjunctive propositions, and logical product instead of adjunctive propositions. It was through these mathematical elements that logic seemed to have been rendered capable of partially reducing mathematics itself to logical forms.

0.7. Logic and mathematics have become competing methodologies of thought. Historically, this was originally inspired by Plato and occurred through the integration in the medieval universities of logic (then called dialectic) into the "trivium" as the group of subjects which later became the "Geisteswissenschaften" or the human or interpretative sciences of "Philosophical Faculties", and of mathematics into the other part of the medieval Philosophical Faculty curriculum, the "quadrivium," the foundation of the modern natural sciences. The two methodological cultures which result from these developments continue to confront each other in a state of mutual incomprehension. The disproportionately lesser weight and influence of "trivial" logic as one among several philosophical disciplines and the overwhelming weight and influence of "quadrivial" mathematics as an independent and important science have immediate consequences for the relationship today between what is called classical logic and mathematical logic.

0.8. The development of mathematical logic (since Boole, Jevons, and Frege, among others) was intended as an attempt to clarify the logical characteristics which are specific to mathematical thought and in that sense it deserves to be continued. But since the founders of mathematical logic and most of its practitioners have themselves been mathematicians, its only result to date has been the application of mathematics to certain logical objects or structures. Whether the basis of mathematical logic is logical or completely non- and extra-logical has therefore remained undecided. Only when this question has been resolved can one decide whether mathematical logic can properly complement, improve, or even replace classical logic, as many mathematical logicians suppose, or whether it is a genuine methodological alternative.

0.8.1. On the answer to this question depend the answers to two others: Whether thought in the "quadrivial" natural sciences according to the mathematical method must proceed in a fundamentally different fashion than in the traditional "trivial" human or interpretative sciences, which proceed according to the non-mathematical methods of classical logic; and whether the human or interpretative sciences can and should be "mathematicized".

0.9. In this book the thesis is put forward that mathematics is an essential part of classical logic and is therefore itself, as a whole, logical. Mathematics is that part of classical logic which, as "dialectic" – that is, the methodological use of contradictions – classical logicians have left to mathematicians. Mathematical logic must therefore exhibit the characteristics of the dialectical method of thought which essentially characterizes mathematics.

0.10. The following are genuine logical elements of the formalism: Concepts, as well as their components, namely, characteristics or intensions, on the one hand, and extensions on the other; logical connectors (junctors or syncategoremata), assertions or propositions (judgments), inferences, as well as their components, premises and conclusions, arguments, theories, and principles (axioms). These concepts have become technical terms used when educated people speak about logic in ordinary language. They describe certain features of linguistic articulations which are distinct from and go beyond what is described by linguistic and grammatical determinations of those articulations. Logic retains the task of defining them as logical concepts and explicating them within a theory by means of propositions and inferences – that is, the task of representing them in a logical formalism.

0.10.1. The meaning of these genuinely logical terms can be demonstrated and their use learned through linguistic examples. But logic is not suited to the task of rendering ungrammatical or even meaningless linguistic material (components of language, which are not themselves language!) meaningful or even logical. What is termed application of logic to linguistic material is always a benevolent interpretation of linguistic expressions as having meaning, with a view to their logical character.

0.10.2. Historically logic has, together with scientific education, rendered the language spoken by educated people itself logical. The result is that many linguistic regions of meaning have been conceptually organized. For example, speakers of a language distinguish between more general and more specific concepts as well as between sentences which make assertions and those which do not. This has given rise to the widespread but false opinion that language is itself, as it were, naturally logical, so that with regard to a certain subject matter, examples drawn from language could themselves show what is actually logical.

0.10.3. From Chinese iconic writing (ideograms) one can learn how all of the features or intensions of a concept can be represented as sensorily perceptible in a single complex sign. The "radical" of such a Chinese ideogram (a simple symbol which also serves for the lexical classification of the signs) generally represents the generic characteristics or features of the concept. The additional component signs (which may themselves function as radicals in other ideograms) represent specific differences. In most cases those components of the sign which include the sound or pronunciation of the Chinese word are also chosen so that they help to express the meaning of the sign as a whole. Similarly, in the pyramidal formalism, all of the characteristics which make up the meaning of the concepts are designated through capital letters in their respective positions in the pyramid. The left-side A signifies the generic intension, whereas the additional letters on the right side signify the specific differences in the lower species and subspecies. See Figures 1 and 2 at 1.4.

0.11. The logician frequently has occasion to consider many of the scientific pronouncements expressed in language as extra- or illogical.

0.12. In this book, the attempt is made to improve logic through some considerations relating to: 1. Development of the classical structure of the conceptual pyramid as the basic logical formalism so that it represents the logical elements and their connections. 2. Supplementation of the theory of concepts by placing in the foreground the formalization of both the intensional and extensional components of concepts. 3. The formalization in pyramidal form of both regular and contradictory ("dialectical") concepts, and a demonstration of the role which induction and deduction play in generating concepts. 4. Distinguishing logical connectors which form expressions from those which form propositions. 5. The arrangement in the pyramid and definition of all logical connectors. 6. A more precise distinction between propositions and definitions, in particular an explanation of the so-called particular and individual propositions as well as of equations as definitions. 7. Formalization of true and false judgments, and those which are true and false at one and the same time, such as the alternative, the contradictory propositions and the probability-assertions. 8. Formalization of the Aristotelian syllogism and of the Stoic inferences and a demonstration of the dialectical nature of probable inferences, and 9. Clarification of the fundamental concepts (or axioms or principles) of logic.

Chapter 1
On Concepts

On intensions and extensions of concepts – Inductive and deductive construction of regular and contradictory (dialectical) and contrary (dispositional) concepts – Their definitions expressed intensionally and extensionally by their position in the pyramid – On complete and incomplete induction in logic and mathematics – Particulars and individuals as quantified concepts insufficiently determined and therefore lacking definition – Multiple inductions of different general concepts from a single individual instance – Deductive construction of contradictory concepts – The creative and heuristic character of contradictory concepts – The formal construction of contradictory concepts – Dispositional concepts and their logical structure – The contradictory structure of number-concepts – The logical concept of prime number and the definition of primes *via* the application of the algorithms of non-primes.

1. Only that can properly be considered a concept which possesses characteristics (intensions) which are distinct and a scope of application (extension) which is clear. The development of true and false insights into the nature of concepts can clearly be seen in representations of "Concept" in the philosophical dictionaries. The fact is that there are no "empty concepts", i.e., concepts which lack either intensions or extensions or both. Nor are "indefinable basic concepts" (understood as categories in Aristotle's Organon and as axioms in mathematics) genuine concepts.
   1.1. Intensions are (recollected) representations of sensory experience. They can be designated, singly or collectively, by words which evoke these representations, as George Berkeley correctly showed. In concepts they are united with extensions. In the pyramidal formalism particular formal intensions are designated by capital letters inscribed in circles, which indicate the positions of the respective concepts within the pyramid. One can also write names or descriptions of the intensions within the circles; this renders the pyramid semi-formal instead of strictly formal.
   1.2. Extensions are also (recollected) representations. They determine a range of experience from which intensions may be established through induction. In the pyramidal formalism they are formally designated by all those

L. Geldsetzer and R.L. Schwartz, *Logical Thinking in the Pyramidal Schema of Concepts: The Logical and Mathematical Elements*, DOI 10.1007/978-94-007-5301-3_1,

subordinate concepts which fall under the position of any given concept, and which evoke that range of experience. The extensions of concepts established through induction always describe "actual" (sensorily experienced) instances, never "possible" (not yet experienced) instances, although the theories of induction dominant since David Hume assume that these latter, too, are included. However, "non-experienced" instances can only be – at most – fantasies. If one experiences and investigates "new" instances, the result can only be either that they are exactly the same as the instances already experienced, or that they are not. But in the latter case such instances don't belong to those of the inductively established concept.

1.2.1. Ernst Cassirer, but also Thomas Aquinas, were of the opinion that the more general a concept, the richer its content. Such a relationship can apply, however, only to the applications (extensions) of concepts. But when one refers to the "content" of concepts, what is meant is their respective intensions. Understood in this literal sense, Cassirer's and Thomas' thesis is false.

1.2.2. The concept of nothing, like the concept of being, is one of the most general inductively established concepts. In accordance with particular sensory experiences, it encompasses concepts which are subordinate species and subspecies. In logic the concept of nothing is generally understood as the meaning and reference of all contradictory concepts and false propositions. But it isn't that at all. Nor does it arise from transforming the logical connector designating negation into a substantive term.

1.3. Regular (non-contradictory) concepts arise from experiences which can be present to the senses all at once and then recollected as such. They differ with respect to their degree of generality, that is, the number of characteristics which they combine. The most general concepts (highest categories) contain at least one characteristic. Subordinate concepts contain increasingly numerous characteristics, but never infinitely many. Scholastic logic wrongly supposed so, as expressed in the Latin maxim "individuum est ineffabile" (it is impossible to articulate all of the characteristics of any individual). So-called proper names or descriptions designate the lowest level concepts in the pyramidal formalism. What distinguishes each of these lowest level concepts from all other concepts is *also* a characteristic or specific difference – in the same way that concepts are always distinct from one another.

1.4. Concepts relating to a sphere of experience group themselves by means of their intensional and extensional interlocking structures within the pyramidal formalism. The conceptual pyramid represents the declining degrees of generality among generic concepts, species, and subspecies or individuals. Generic concepts have either two (dihaeretic, that is mutually exclusive) or several (multiple) subordinate concepts or species. Only a conceptual pyramid which consists exclusively of regular, i.e. non-contradictory concepts can itself be completely free of contradictions. Such a conceptual pyramid embodies the logical formalism of a completely consistent ("coherent") theory covering a certain domain of experience (Figs. 1.1 and 1.2).

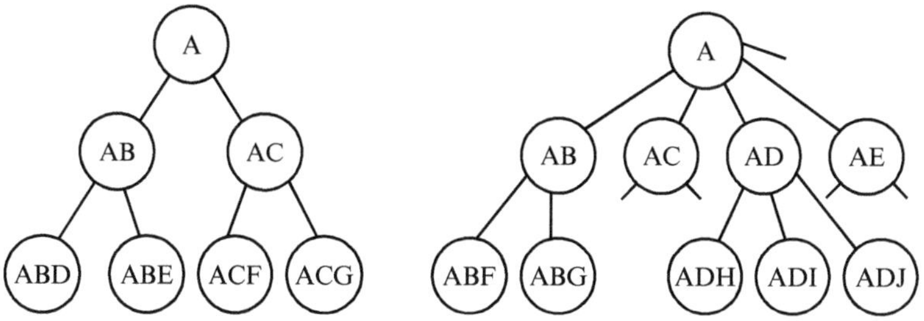

Figs. 1.1 and 1.2 Extensional-dihaeretic and extensional-multiple conceptual pyramid with inscribed intensions

1.5. The fact that the pyramidal formalism shows differing degrees of generality doesn't mean that more general concepts are "higher" than those which are less general. Porphyrian trees were drawn in antiquity and the middle ages to show the "higher" degree of generality at the bottom, with less general concepts represented as higher branches (Compare Porphyry's Introduction to Aristotle's Organon). Here again general concepts are not to be understood as "deeper" or "lower." Both ways of representing them are unavoidable optical deceptions of the formalism. Throughout the history of philosophy these have led to significant mistakes in the ontological evaluation of general concepts (universals). Hegel remarked – correctly – that one penetrates into concepts by means of intensions and that, consequently, general conceptual content is always and only implicit in concrete particulars, and not located anywhere else.

1.6. A regular conceptual pyramid shows what a logical induction (Greek: epagogé) – the process of abstracting to and from concepts – is. Lowest level concepts are compared and reviewed in order to determine which of their characteristics are the same, that is, represent identities, and what their specific differences are. Identical characteristics are designated by the same letters. Common characteristics are combined in so-called higher level or more general concepts and the same process is repeated. For a given range of experience, a highest level generic concept is finally constructed, which contains at least one "generic" characteristic which is common to all and identical in all of the lower-level concepts within that range.

1.6.1. A logical induction is always complete. According to William of Ockham, a single individual is a sufficient basis for an induction. And according to Francis Bacon, the number of examples used as instances has no logical relevance, since in logic one does not count. It is apparent that a given characteristic of a concept, which has been inductively established from an example, fixes once and for all whether as yet "unknown instances" display that characteristic or not and accordingly whether or not they fall within the extension of the inductively established concept.

1.6.2. The "black Australian swan" which was discovered during the lifetime of John Stuart Mill didn't call into question the complete induction of the concept of the swan, but the discovery did suggest that the color white was not a generic characteristic of that concept and had to be eliminated from the intensional characteristics of the concept "swan". This did lead – inductively – to a new division of species of swans according to color.

1.7. The so-called complete induction in mathematics (for example, for the mathematical definition of the concept of number) is incomplete and for that reason not logical. In mathematics one counts, but no one has ever counted "all numbers". Usually, following B. Russell, two concepts are posited, a "first element" and a "successor". But it is apparent that one cannot inductively establish a generic concept of number from these two concepts, i.e., one that contains a generic characteristic of both. Mathematics still lacks a logical conception of number. In this book we propose one.

1.8. Induction is a method of constructing concepts by means of definitions, not of forming propositions which express assertions. Certainly it is not a procedure for the formulation of causal laws, as J. St. Mill supposed in his "System of Logic" of 1843. Induction and inductively established concepts have no truth value.

1.8.1. What is termed the "incompleteness" of induction properly refers to the fact that many concepts are not sufficiently determined and lack internal structure. A concept can only be precisely determined, that is, defined, through the complete specification of its intensions and extensions.

1.8.2. To the extent that a concept lacks such complete specification and is therefore not fully determined, it is either not clear, or not distinct, or both. Logic is often thought to operate fundamentally with concepts which are clear and distinct, that is, completely determined. But that is one of the principal errors which infect traditional and current conceptions of logic.

1.8.3. Concepts which are insufficiently determined can be more precisely or completely determined by means of definitions. Definitions describe concepts with insufficiently determined extensions by a (more precise) specification of their intensions and concepts with insufficiently defined intensions by a (more precise) specification of their extensions. As a rule, definitions represent a concept by means of two equivalent expressions. This definitional equivalence is reversible (commutative).

1.8.4. Particular and individual concepts are examples of concepts which are insufficiently determined. "Some living things" or "a living thing" does not specify which species or which individual thing is meant.

The concept only becomes clear and distinct when it is supplemented through an equivalence with an intension such as "animal" or "a dog." The precise formalization of the equivalence in this instance is: "Some living things = animals = not plants" or "a living thing = (this) dog."

1.8.5. W. V. O. Quine's explication of "gavagai" (from an imaginary language) shows that he did not understand the structure of a scientifically defined concept. The consequences that he drew from that explication for a holistic theory of science and the indeterminacy of translation of concepts and theories are therefore themselves in error.

1.9. The procedure of induction itself does not in any way determine the direction in which one abstracts from particular concepts to more general – higher level – ones. One may freely select among the characteristics of a given empirical instance that characteristic which is to be used for the induction. That choice determines an inductive framework within which other characteristics of the same lower level concepts are assigned to them as specific differences.

1.9.1. It is their different inductive frameworks which distinguish different theories directed at the same sphere of experience. Epicurus appears to have been the first to notice this. G. W. Leibniz in his Monadology and J. M. Chladenius in his hermeneutics developed on that basis the theory of "viewpoints". It is by means of an inductive framework that theories subordinate data to a certain point of view and arrange them by quantifying extensions according to species ("some") and individuals ("a"). As an example, take the characteristics of the individual Socrates. These may be said to properly describe the individual Socrates without regard to the sequence in which they are used. Each of them may supply an inductive framework and each permits different definitions of "Socrates" (Fig. 1.3).

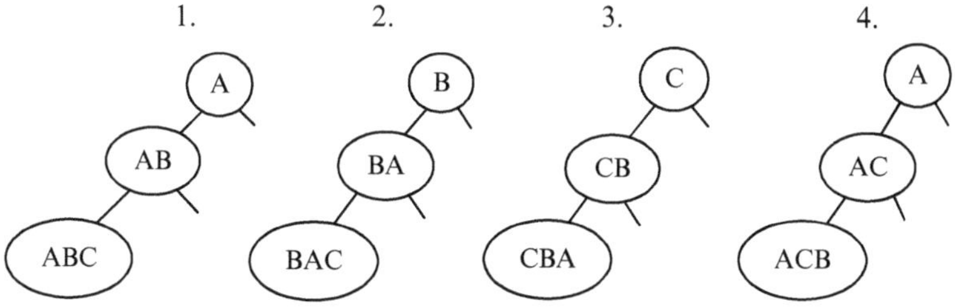

Fig. 1.3 Four possibilities of complete induction from one individual instance

A, B and C are the inductive frames established from the three characteristics: Greek (A), Philosophic (B), and Socratic (C), of the person called Socrates.

Read the definition of the Individual (Socrates) at the base of the pyramid by using the quantified upper concepts, beginning with the genus (the inductive frame concept):

I. ABC = a Greek of philosophic kind with the specific denomination Socrates.
II. BAC = a Philosopher of the Greek kind with the specific denomination Socrates.
III. CBA = a Socratic of the philosophic kind with the specification of a Greek.
IV. ACB = a Greek of the Socratic kind with the specification of a philosopher.

1.9.2. This method of induction has to be distinguished from mere analysis of characteristics of terms and expressions in literary and scientific texts which is used in bibliographic cross-checking. Catch-words designate the places in texts where similar concepts may be found, but they do not designate their exact meanings.

1.9.3. Induction (and likewise deduction) has also to be distinguished from mere classification of themes or topics which is used in rhetorical presentations of ideas, concepts or parts of theories.

1.10. A regular logical deduction presupposes the inductive establishment of an inductive framework with a most general concept which is thereby properly defined. This inductive framework contains the generic characteristic. In a deduction the other characteristics which have been established through induction are distributed among the subordinate conceptual positions as specific characteristics and supply the definitions of these subordinate concepts. The first figure of the formalism set forth in Fig. 1.3 can be deductively read as "some A = AB," "a (definite) A = ABC."

1.10.1. Such deductions occur by means of what are traditionally called particular and individual propositions (subalternations). These, however, are not propositions with truth values, as logicians have traditionally assumed, but definitions. In fact, these definitions are conventional assignments of intensions (as specific differences) to conceptual positions under a general, high level concept (within an inductive framework) which are – initially – only quantified in extensional terms and consequently insufficiently determined. For these reasons, definitions are properly formulated as equivalences, employing the equals sign or the expression "in other words" rather than "is", the copula.

1.10.2. It should be apparent that such a "formal conceptual analysis" (see 0.1.3.) can also be logically performed by pointing out those characteristics of high-level concepts which are also contained in the basis concepts given in the data of a particular science.

1.10.3. When dihaeretic (mutually exclusive) species concepts are defined the result is pairs of positive and negative definitions. One can read off from the figures set forth above: (1) Some Greeks = philosophers/some Greeks = non-philosophers; (2) Some philosophers = Greeks/Some philosophers = non-Greeks; (3) Some Socratics = philosophers/some Socratics = non-philosophers; (4) Some Greeks = Socratics/some Greeks = non-Socratics. Definitions (as supposed particular and individual propositions) can be easily recognized by means of a "cross-check" to determine the existence of a coordinate species concept (or in appropriate cases a coordinate individual), which are here designated by negations. (But the latter are not noted in the pyramid above.).

1.10.4. As a rule, individuals are multiple coordinate species concepts at the lowest level. They are most often defined solely in a positive fashion; for example, "a Greek = Socrates the philosopher." The cross-check distinguishes the individual from all other individuals taken collectively, for example, "no other than …".

1.10.5. Kant famously proposed a refutation of Anselm of Canterbury's (or, as he called it: the "Cartesian") proof of the existence of God. Kant wanted to show that one could not properly deduce the existence of an individual falling under a general concept from the general concept itself. He therefore concluded *paradigmatically* from the concept "thaler" (an eighteenth-century German coin) that "there were no thalers in his pocket." In order to accurately determine that fact, however, he required at a minimum a clear and distinct concept of a thaler. And from that concept he should have more precisely deduced that there were "some thalers in pockets" and "some thalers not in pockets." Only having first drawn this inference could he then properly further deduce that, if there were no thalers to be found in the individual Kantian pocket (or in the "state of his personal property", as he said), at least one thaler must be found, if not in other pockets, than somewhere else entirely.

1.10.6. In the case of numerically quantified concepts (metric concepts, particularly in physics) the logical particularization expressed by "some" is replaced by numerical values. In such cases, the unit of measurement serves the purpose of logical individuation. A non-quantified metric concept has the logical quantifier "all," which is implicitly understood in the case of such general concepts and is therefore not specifically mentioned or designated. Cassirer's "functional concepts" are meant to be numerically quantified concepts.

1.11. In mathematical logic particularization and individuation are usually combined in the expression "at least one …" This is considered an advance in precision; in fact it is the reverse.

1.11.1. In mathematical logic, it is customary to combine the existential logical connector ("there is") with particularization and individuation.

This usage reflects the false assumption that definitions are assertions with truth values. A proposition such as "there is at least one x such that x = (f) y" simply expresses in a pleonastic manner that the concept x, which is not fully determined, is to be understood as a specification of the higher level concept y (an intensional definition) and that x has quantifiable extensions.

1.12. The fusion ("synthesis") of the intensions of specific concepts with those of more general concepts occurs at the linguistic level through the formation of such modifying expressions as "the house of the guests" or through direct combination of the words in question, with or without a hyphen ("guest-house" or "guesthouse"). In the example the concept "house" is specified through the concept "guest." The concept "house" is named together with one of the concepts that is subordinate to it, so that the generic characteristic of the subordinate concept is exhibited at the same time as the specific difference of the combined concept, "guesthouse." The relationship between the two concepts is transitive, that is, not reversible (compare: house-guest).

1.12.1. When the intensions of separate and complete concepts – which are not themselves species-concepts with a generic concept in common – are fused in a third concept, we may call the result a "*product concept*." Product concepts are also expressed through word-combinations with or without hyphens. In such cases the fusion is commutative. This procedure is at the root of the formation of many concepts used in physics, by means of mathematical products (multiplication), for example: "mass · acceleration = accelerated mass = energy." Multiplication should be understood not as the linking of concepts by means of a logical connector, but as an immediate fusion of concepts.

1.12.2. The logical concept of the concept is itself a product of the concepts "intension" and "extension". Concept = intension · extension = extensionalized intension = intensionalized extension.

1.12.3. In relational logic, so-called relational concepts are introduced as indeterminate predicates with two or more places for determinate other concepts. But in fact they are expressions in which several concepts are linked together by means of either logical connectors of the type which do not form assertive propositions, or linguistic particles which have a connective function. Otherwise, however, the concept "relation" and its species and subspecies can be defined in a conceptual pyramid in the same manner as any other concept. Certain mathematical relations (for example: larger than, smaller than, between, antecedent, successor) have become objects of study in the logic of relations. What this fact shows, however, is that mathematics does not yet have at its command a logical theory of mathematical relations.

1.13. In mathematics, mathematical deductions are considered a method of proof. It is said (following Euclid) that by this means the truth of theorems (propositions considered as the teaching-units of a discipline or theory) can be demonstrated by deriving them from axioms and from other theorems. However, to the extent that such "proofs" involve the representation of equations and their successive modifications up to the so-called solution, what they show is actually only that all definitions which occur and are used in the case of the proof are ostensibly well-formed. In the case of arithmetical problems a solution consists in this: that, by means of equivalence, a mathematical expression is defined as equivalent to a numerical value (and the reverse).

1.14. As a rule, contradictory concepts (contradictions "in terms" or "in adiecto") are established deductively. They fuse inductively established regular dihaeretic concepts (which stand in a relation of complete disjunction to each other) in a single concept which contains, at the same time and in common, specific differences as characteristics which "exclude" one another. For that reason, such concepts can be constituted only out of heterogeneous recollections and by means of fantasy (speculation) rather than directly from sensory perceptions, which are then fixed in memory. One can, for example, not perceive a living thing as living and as dead at the same time; but one can certainly recall both such conditions and unite them in the concept of "mortality". That is why logicians have traditionally asserted that there is in reality "nothing" which corresponds to contradictory concepts. But what does in fact correspond to them are segments of reality which do not belong to a common experiential context. Were it otherwise, one would be unable to represent anything at all in connection with such concepts. Contradictory concepts are the conceptual building blocks of "possible worlds."

1.14.1. Hegel's dialectic distinguishes itself through this: That he defines all sensory and intellectual activities on the one hand, and practical activities on the other, as contradictions in terms. Thus, Hegel defines "experience" as a fusion of object and subject (or thing and ego), sensory "perception and deception" as a fusion of the general and the individual (or of the "this" and the "opinion"), comprehension (through concepts) as a fusion of that which is given to the senses and of the understanding's apprehension of essences; activity directed toward satisfaction of desires or will as a fusion of the "unhappy or slavish" consciousness with the "lordly or free" consciousness; creative activity (resulting in works of art, or historically significant actions) as a fusion of personality and objective accomplishment or of institutions and values. The explication of these forms of action must itself involve contradictory assertions. Hegel has, however, embedded these dialectical passages in a context which is free of contradictions.

1.14.2. Contradictory concepts cannot be negated because the specific differences which they contain already negate each other. Negation results only in the exchange of these meanings, and the meaning of the concept as a whole remains the same. Nevertheless, concepts which are contradictory but not recognized as such are frequently negated. That necessarily results in further contradictions when the concepts are employed in expressions and propositions.

1.14.3. The most prominent contradictory concept in logic and in many of the sciences is that of possibility. It is formed by fusing the concepts of "being" and "nothing," each of which can be inductively established (see: 1.2.2). Therefore "possibility" is certainly not the common generic concept of "being" and "nothing." If it were, it would be formed by abstracting from – disregarding – the specific characteristics of "being" and of "nothing" (or "not-being"). But in fact the concept of possibility encompasses the characteristics of both of these concepts – characteristics which exclude each other. Everything possible is therefore both being and nothing. "Possible worlds" are not worlds in being, but to the extent they have been conceived and constructed (particularly by logicians) they are also not nothing.

1.14.4. The concept of possibility is – as is true of contradictory concepts generally – not capable of being negated. If it were, "impossible" would be understood as both not-being (the initial concept of being replaced by its negation) and at the same time not-nothing. But in that event the first concept would mean "nothing" and the second "being." In his theory of modalities, Aristotle defined the "impossible" as the contradictory itself. What this shows, however, is only that he failed to recognize the contradictory character of the "possible."

1.14.5. Contradictory concepts are employed so frequently in ordinary language that they are customarily considered regular concepts. For example, we referred above to the concept of "mortality," which is traditionally used as a regular conceptual predicate in propositions concerning human beings. ("All men are mortal"). In fact, however, human beings are only either living or dead. The "mortal Socrates" is, we know, long since dead. At the same time, however, as we also know, he is very much alive in the memory of western culture. His "mortality" can therefore mean only that he is both: dead and alive.

1.14.6. As speculative models for mathematics, art and technology, contradictory concepts are creative, because they extend beyond our sensory experience. In Christian theology they take the form of articles of faith (dogmas). In scientific research they are heuristically fruitful and therefore indispensable for directing and fixing attention. In art and technology they are preconceptions of realisable creations. In ordinary language, they are considered expressions for "impossibilities" (oxymorons such as "bright darkness" or "curved straight line"). When used in arguments, they give rise to double and ambiguous meanings, each of which can be used to call the other into question, as in the case of paradoxes. The formalism of these contradictions provides a foundation for a logical theory of creativity.

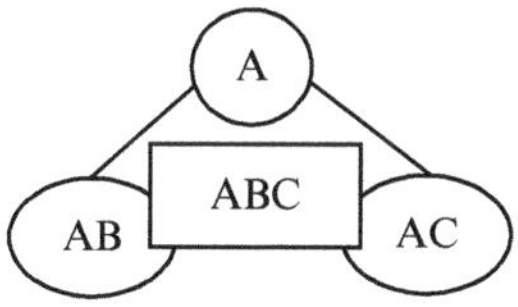

Fig. 1.4 Establishing contradictory concepts by fusion of dihaeretic species (Biological example: A = organism; AB = living organism; AC = dead organism; ABC = mortal organism; Logical example: A = value; AB = truth value; AC = falsity value; ABC=truth-falsity value (or: the Third) (see also 7.3 and 7.5))

1.14.7. Creativity is generally considered a phenomenon which resists theoretical comprehension. This opinion derives, perhaps, from theological myths of creation and a secular cult of genius. Yet the simplest "creation" consists of amalgamating two inductively established concepts from heterogeneous spheres of experience into concepts which are contradictory or dispositional or both, and which in this unified form don't occur in experience. This act of unification or synthesis can only be achieved through the exercise of human memory and fantasy. For example, the concepts of horse and bird can be fused to form the dispositional concept of "Pegasus," as the fantasy of artists has long since shown. As a heuristic concept it can serve to inspire research into the question of whether winged horses actually exist, a type of research which was often undertaken in antiquity. Employed in genetic technology as a model for breeding, the same concept could lead to the production of similar animals. Similar results have already been achieved in the breeding of "tomato-potatoes" or "potato-tomatoes" and many other hybrid living beings. Such a living being might be discovered or produced. In that case, its concept would lose its contradictory character and, on the basis of experience with the new concept, which could be inductively established, become a regular concept. The new concept should then be coordinated with one of the two initial concepts as a subordinate species. For example, in the case of Pegasus, a Pegasus which had been genetically bred would be either a horse-like bird or a bird-like horse. The same analysis applies to all technical products and beyond them to the structures of "possible worlds (Fig. 1.5)."

1.14.8. Contradictory concepts can be inscribed in each location in a conceptual pyramid between two dihaeretic (completely disjunctive) concepts as their amalgamated product. They contain within themselves the specific differences of each of these initial concepts – specific differences which exclude i.e. negate each other. It is this which distinguishes them from regular general concepts, which do

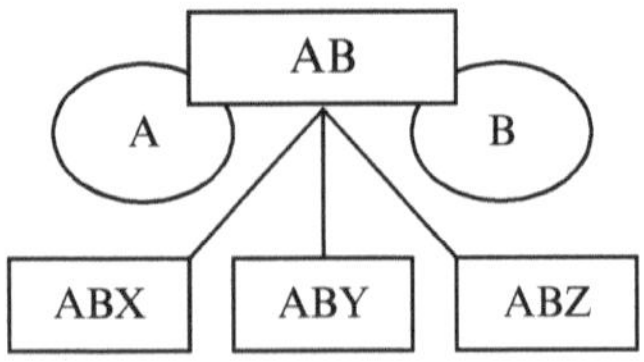

Fig. 1.5 Deduction of possible worlds (A = Being, B = Non-Being, AB = Possibility, ABX = possible world X (f. i. the Past), ABY = possible world Y (f. i. actual Reality), ABZ = possible world Z (f. i. the Future))

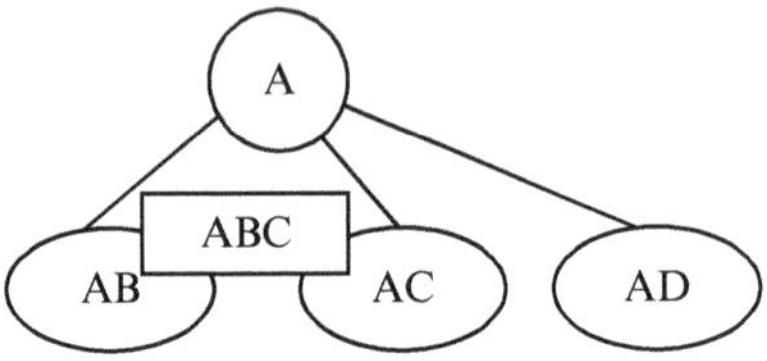

Fig. 1.6 Establishing contrary or dispositional concepts from multiple species (Example: A = state of aggregation, AB = solid state, AC = liquid state, AD = gaseous state, ABC = meltable (dispositional) state)

not contain such specific differences. They possess no unique extension which belongs to them alone – and that is why they are thought to have no basis in reality. Precisely understood, however, the contradictory concept encompasses – as does the regular general concept – the extensions of the initial concepts which have been unified within it. That is why many contradictory concepts are confused with regular general concepts. As we've said, for example, the concept of possibility is formed by means of the contradictory fusion of the concepts "being" and "nothing." In modal logic, however, "possibility" is customarily considered to be the most general, highest level, and regular concept. And modal logicians therefore consider the actual "existing" world and the "non-existing" worlds to be species or realms of "possible worlds" (Fig. 1.5).

1.14.9. Irregular so-called contrary (in distinction from contradictory) concepts may be inscribed between two concepts in a series of multiple species-concepts. The extension of such a concept consists of the extensions of the two concepts which it unites. These are the so-called *dispositional concepts* which are employed with particular frequency in physics and the natural sciences and psychology (powers, capacities, forces, dispositions) (Fig. 1.6).

1.15. The concept of number is a contradictory concept. It can be deduced from the logical connectors (positive quantifiers) which are used for quantification and so defined. As a generic concept it encompasses species and subspecies-concepts with their respective numeric individuals ("numbers"). As a logical matter, the numeric individuals do not show themselves to be well-ordered in the sense of a sequence. This lack of logical sequence obtains with respect to the so-called cardinal numbers. Their "magnitude" or "size" must be learned through the immediate perception of easily grasped groups. In this way, mostly by comparison with fingers or toes, or both, one learns certain groups, such as pairs, triples, fours and fives, dozens, and in some cases twenties. Most languages have their own elementary numerical designations for such groups.

1.15.1. The so-called ordinal numbers (the first, the second, etc.) are first produced within the particular numeric species and subspecies through the application of particular arithmetic ordering procedures (for example, the formation of sums, from an initial number and an added unit). The procedure itself perhaps derives from the experience of a sequence of births in a family, which is reflected in many customary Roman and Japanese children's names. These names usually are situated within the range of the total number of children born to a single mother. Social ranks may also have played a role in giving rise to the designations of ordinal numbers (for example, the prince as the (Latin) "princeps" = the "first head"). Ordinal numbers possess their own dialectic. They essentially designate only a unit which supervenes on a given number (set) of objects. But at the same time the initial number must also be thought of. This is the basis of the customary (Russellian) mathematical definition of number as "initial element and successor." In Euclid's time, "one" was considered the initial element, so that "two" was the first number which was actually defined. Now, however, zero is posited as the initial element, so that "one" can be obtained as the first defined number. (This obviously amounts to a "creatio ex nihilo" (creation out of nothing) of numbers.)

1.15.2. The logical concept of number arises from the fusion of quantifying logical connectors which stand in a dihaeretic (mutually exclusive) relation to each other. They are the quantifying logical connectors which signal individualization ("a" or "one of") and universalization ("all"). One should here bear in mind the sensory representations which are effectively united when we "comprehend" contradictory concepts. In experience, we can fix our attention sequentially either on "an" object or on "all" objects within a given range of perception, but only in recollection or in fantasy can we simultaneously represent to ourselves "one and all." That is, however, precisely what is required in a representation of number, namely that one thinks "all (the totality) as one" or, in reverse, "one as (a totality) of all."

1.15.3. The mathematical concept of number takes the place of the quantifying logical connector which expresses particularization ("some"). In logic "some" remains an independent quantifier. In all quantifying sciences, it is replaced by numerical values, each of which fuses logical unity (individuality) and logical universality. This contradictory fusion of the logical connectors totality and unity defines the concept of number as "(individual) unity of totalities" or "totality of units", or more succinctly as "all-one(s)". These fused intensions of the general concept of number remain present as general characteristics and components of the meaning of all concepts of different kinds of numbers and of the (individual) particular numbers.

1.15.4. As long as arithmetic was understood as the quantification of geometrical structures, as it was by Euclid (an understanding which persisted well into the time of Descartes), arithmetical concepts had geometric intensions and numerical extensions. That is why the "one" was understood not as a number but as a logical unity or unit of geometrical objects. In the same way, the infinite could not be a number, it could only be a logical totality (the all of). In the place of zero there was the logical "none" (no one of) and negative numbers were still simply negated quantitative expressions. This explains the fact that the Indian-Arabic zero (as representing "none") was not absent from European calculation.

1.15.5. In the "quadrivial" physical sciences arithmetic quantification is applied to geometrically constituted or modeled natural objects. Everything measurable must first be given geometric form.

1.15.6. Modern arithmetic makes the quantifiers themselves into the intensions of arithmetic concepts and quantifies them at the same time. That is the basis of their dialectical character. The meaning (intension) of a number becomes its extension. And the extension of a number becomes its meaning.

1.15.7. In mathematics itself, the general concept of number is what is called "the natural number" (but naturally, in mathematics nothing at all is "natural".) The initial concepts which are fused in the general concept of number ("one" and "all", or "individual unity" and "totality") stand logically in a relationship of complete disjunction to each other. As a logical matter, therefore, "(individual) unity is not totality" and "totality is not unity." In mathematical logic and in arithmetic, however, "each totality is at the same time a (individual) unity (of totalities and/or units) and each (individual) unity is at the same time a totality (of totalities and/or units)". In this sense, the concept of number is defined as "simultaneously logical unity and logical totality."

1.15.8. Mathematical logic discusses the attributes of the concept of number in the terminology of set theory. "Set" is, from a logical viewpoint, simply another word for the logical connector which signals particularization. For that reason, "set", as understood in classical logic, means the same as "some." Whoever, in logic, proceeds on the understanding that a "set of things, or some things" is neither "all things," "one thing," nor "no thing," is still, even today, thinking logically. In order to think mathematically and according to mathematical logic, however, one must adapt oneself to the fact that the concept of set used in mathematical logic can assume the meaning of all of the quantifiers used in classical logic. Next to the ordinary "set" (which still means "some"), there is the "universal set," the "unitary set" (with only one element as content of the set), and the "empty set" (the zero set, with no content). And beyond this, the concept of a set is used "reflexively," that is, mathematical sets are occasionally said to "contain themselves." By so doing, however, one set would have to be – simultaneously – "two sets." Such reflexivity is excluded for regular logical concepts. One easily sees: It is not the concept of the set which explains number, but the reverse. "Set" is understood according to the meaning of number. The analysis of the concept of the set in mathematical logic (which is itself considered as the general concept of number) may be said to have sufficiently laid bare its contradictory structure. This has been achieved through work – now lasting over a hundred years – on the paradoxes which can be formulated using the concept of set.

1.15.9. Apart from numbers, there are a number of other concepts in mathematics which do not deny their logical origin, although these are not usually recognized. When one speaks of large or small numbers, it is understood logically from experience that in the case of quantifiable objects, "all objects" usually form a larger pile than "some" or "one." And if one divides one such object into pieces (like the number one into fractions) they can become so tiny that one can't see them anymore, and they apparently vanish. This sensory experience serves as a didactic instrument when arithmetic is taught. It also serves – to a greater extent than is acknowledged – as a metaphorical model material for the purely "rational" treatment of mathematical objects. Then, starting from any given number, one quickly attains the "infinitely large" ("infinite") and the infinitely small ("infinitesimal") and acts as if one had to investigate their behavior with intellectual telescopes and microscopes. However, if one examines them with the aid of the (Platonic) "eye of the intellect" for which there can be nothing large or small, and therefore no "limes values" and certainly nothing "without end" or infinite (Greek: apeiron) one notices how dialectical creativity works itself out mathematically in the case of these concepts as well. The so-called infinitely large is and remains a logical totality, applied to all numbers. Since totality is a logical concept, it cannot itself also be a number. As adapted in arithmetic, however, totality as the infinitely large is understood to be both number and non-number at the same time (Georg Cantor's theory of magnitudes ("Mächtigkeiten") creatively develops this point further). Similarly, the infinitely small or the infinitesimal, which is "less than the value of any number" is a logical unit, and therefore not a number; but in arithmetic it is required to be both a number and not a number at the same time. This is the subject of infinitesimal mathematics and the theory of non-standard numbers. The logical "none" is also present in arithmetic. It becomes zero, which therefore must also be both number and non-number. In the future, as well, there is no conceivable limit to the deduction and definition of further "units" in mathematics. Methods of calculating with letters have introduced logical symbols for concepts, which may indifferently represent either numbers or other units. Indeed, letters functioning as variables were introduced into modern arithmetic as logical symbols of concepts in order to disguise precisely this distinction between numbers and concepts.

1.16. The concepts of specific kinds or species of numbers are here dialectically deduced, according to the pattern of regular dihaeretic definitions of concepts. This doesn't at all reflect their historical invention, discovery, or construction, as is apparent from their often peculiar names. In their turn, dialectical (contradictory) concepts of numbers can be defined from the dihaeretic species and subspecies-concepts of particular kinds of numbers. A three-level pyramid of these concepts is given here. The formalism of the conceptual pyramid is used here in a semi-formal mode. In other words, the linguistic designations (terms) replace letters standing for intensions. As a rule, these express in words only the specific characteristics of the concept in question. Where there is a contradictory fusion of dihaeretic coordinate concepts the result is two specific characteristics which negate each other. The generic characteristics can be read off from their respective higher level concepts. The deduction can proceed creatively and arbitrarily further in investigations of the theory of numbers or their technical construction. And it will be able to do so even more effectively to the extent that this insight becomes known in arithmetic: That its conceptual creations owe their origin essentially to the deductive and constructive unfolding of dialectical logic (Fig. 1.7).

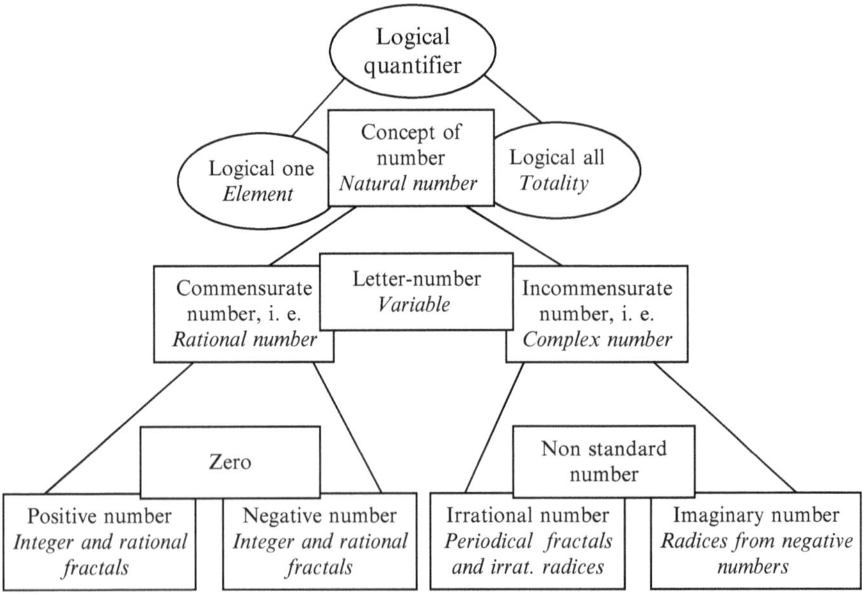

Fig. 1.7 Deduktion of the concept of numbers by fusion of the general and the individual logical quantor

1.16.1. *Legend:* The logical quantifier itself is defined as a logical connector which forms expressions. (See 2.7 below.) It is the general concept for a level of multiple specific concepts with three positive members, namely the individualizing quantifier ("a", "one" or unity), the particularizing quantifier ("some" or multiplicity or logical set), and the universalizing quantifier ("all" or totality). These remain as logical connectors in the vocabulary of mathematics. One speaks, for example, of one, some, and/or all numbers. For the general concept of number, the logical "some" is replaced by the concept of number, which is formed by fusing the logical unity and logical totality. The arithmetic unity thereby becomes the number one and the arithmetic totality an infinite number. Each "natural number" is an example of this general concept of number and defined by means of it. The natural numbers can be divided into commensurate and incommensurate numbers. The basis of the commensurate numbers is the arithmetic one as logical unity or as specific difference. The basis of the incommensurate numbers is some other logical unity (not one of the commensurate unities). The rational numbers are examples of the commensurate numbers, the complex numbers examples of the incommensurate numbers. The concept of the numeric variable (use of a letter as a variable) results from the fusion of commensurate and incommensurate numbers. Letter variables remove the distinction, that is, render it irrelevant for purposes of calculation. An example is the number π (Pi) for the incommensurable relationship between the units of the circumference and the diameter of a circle. From the perspective of geometry and sensory perception, there are different units for curved and straight lines. The commensurate numbers (with the numerical unit one) can be divided into positive and negative numbers and these further into integers and rational fractals (the latter division is only suggested in the diagram). The fusion of each positive number with its negative counterpart leads to the concept of zero (and there must, consequently, be as many zeros as there are positive or negative numbers. Arithmetic still lacks a theory of zeros, just as ontology still lacks a theory of nothingness). The incommensurate numbers can be divided into irrational and imaginary numbers. The specific characteristic of the former is that they unite in the complex a numerical unity, which rests on the arithmetic unit "one," and a logical "infinite" unity, which is based on the logical totality (the latter is designated not through number signs, but through a linguistic convention ("...") indicating "and so forth"). The imaginary numbers combine with arithmetic units an "imaginary unit" ("$i=\sqrt{-1}$") in the complex, which is defined as an exception from the basic rule of the formation of products. The non-standard numbers (A. Robinson) fuse these two specific differences in their incommensurate complexes, so that an infinitesimal "imaginary" component is attributed to each standard number.

1.16.2. Not all types of numbers are shown in the pyramid of number concepts. How further number concepts could be constructed by specification or fusion is partly trivial, partly a subject for further research. The usual definition of a kind of numbers consists in its algorithm, which indicates the reckoning mode (sum, product, partition etc.) "recursively" applied on a basic unit to produce the series of "all" numbers having the same specific character.

1.16.3. The division of the (positive and negative) integers in even and uneven numbers appears trivial, but is not, because of the numerous possible definitions of integers by means of the arithmetic operations (addition, subtraction, multiplication, quotient). The fact that even and odd numbers alternate in the decimal number sequence when one is engaged in simple counting – learned by memorization – only appears trivial when one takes for granted that the numbers are well-ordered and arranged in the decimal system. In mathematics it has been decided to define even and odd numbers by means of their divisibility. It should be noted that Euclid distinguished between division into halves and other forms of division. The even numbers are, whereas the odd numbers are not, divisible into halves without remainder.

1.16.4. Even numbers can be dihaeretically divided into the species of the powers of 2 and into the non-powers of 2. The powers of 2 are defined as 2^x=2,4,8,16,32,64… (for x=1,2,3,4,5,6…). The non-powers of 2 are defined as products of all even and odd numbers larger than 1, that is x·y=#,6,#,10,12,14,#,18,20,22,24,26,28,30,#,34,36,38,40,42,44,46,48,50,52,54,56,58,60,62,#,… (for x=2,4,6,8,10,12,14,16,18,20 …; y=3,5,7,9,11,13,15,17,19 …). These products may be calculated with supercomputers up to any desired magnitude and ordered in the decimal system. As one sees, the lacunae ("#") in the decimal series of the non-powers of 2 indicate the powers of 2 themselves. This suggests a method for calculating the prime numbers.

1.17. The prime numbers are situated as it were irregularly between the even and odd numbers in their decimal order. They are said to be divisible only through themselves and one, while all other natural numbers are "also" divisible through themselves and one. In that sense they represent a selection from the even and odd numbers which constitutes their notorious incalculable character. This definition fits the number two, but this has remained to date the only even prime number, to the extent that the divisibility of "big" numbers has been investigated at all. Beyond that, each larger even number is divisible into halves, and for that reason alone cannot be prime. It is now taken for granted that the number one is not prime, although it fully satisfies the definition given above. These irregularities are certainly mathematically desirable and give the prime numbers their notoriously exceptional position among the numbers (including the number two and excluding the number one), but they are not logical, and they signal the distinction between the logical and the mathematical construction of number concepts. The logical definition of prime numbers must include the number one and exclude the number two. Beyond that it is logically necessary to distinguish divisibility into halves from other forms of division. If, therefore, the logical definition of the odd numbers is "all integers not divisible into halves" then the odd numbers can be divided dihaeretically into the two species of the odd prime numbers and odd non-prime numbers and can be inserted in appropriate locations in the pyramid.

1.17.1. Accordingly, the logical definition of prime number reads: "Prime numbers = all integers which are divisible (other than by halves) only through themselves". This definition includes the number one and excludes the number two. Prime numbers therefore form the dihaeretic coordinate-species of the non-primes, which are "odd integers which are divisible (other than by halves) not only through themselves".

1.17.2. The non-primes might be defined as "odd integers not only divisible through themselves (other than by halves)". Or more precisely: "Non-primes" = all products of odd integers larger than 1. Mathematically: $x \cdot y = 9,15,21,25,27,33,35\ldots$ (for $x = 3,5,7,9\ldots$; $y = 3,5,7,9\ldots$). The (odd) non-primes can be calculated with modern computers up to optional largeness and ordered in a decimal series. And here also the lacunae in the decimal series indicate the positions of the primes.

1.17.3. The mathematical definition and therefore calculability of the primes follows from these conditions: namely that they are exclusively odd integers and that their co-species, the non-primes, are calculable. The magnitude of the difference between neighboring non-primes in the decimal series indicates the number of the primes between them. Mathematically: "$x - y = z$ (for: x = larger non-prime; y = smaller neighboring non-prime; z = an even number). The formula "$(z-2)/2$" yields the number of the primes between the neighboring non-primes (for $z = 2$: no prime number; for $z = 4$: one prime; for $z = 6$: twin-primes; for $z = 8$: three primes, etc.)

1.17.4. The prime numbers are therefore calculated by reiterated additions of 2 to the next smaller non-prime and/or by reiterated subtractions of 2 from the next larger non-prime.

1.17.5. The non-primes might be divided into multiple sub-species. This is done by their definitions as sums of a basic unity and an addendum with a constant structure. We define three sub-species: 1. Base 9 with the addendum of the products of 6: $9 + 6x = 9,15,21,27,33,39,45,51,57,63\ldots$ (for $x = 0,1,2,3\ldots$). 2. Base 15 with the addendum of the products of 10: $15 + 10x = 15,25,35,45,55,65\ldots$. 3. Base 21 with the addendum of the products of 14: $21 + 14x = 21,35,49,63,77,91,$

105... Generally: Base + addendum x · y (for y = 0,1,2,3...). This definition of the sub-species of the non-primes reveals the otherwise incalculable position of the non-primes in the decimal series.

1.17.6. It also appears that the primes should be calculable in decimal series of their sub-species. Their notorious irregularity is probably also founded in the fact that they follow one another in a regular way as sums of different bases and different addendums.

1.18. Knowledge of numbers' respective places in the decimal system is not sufficient for knowledge of the numbers themselves. Knowing everything about a number would require that one knew and actually had present in consciousness all of the ways in which it could be defined through calculations. This more precise knowledge is most likely to be achieved in the case of the smaller prime numbers, because they are not defined through quotients and products. "Plato's number" was famous in antiquity because of the extremely numerous ways in which it is divisible. In 1823, J. F. Fries demonstrated what number it is.

Chapter 2
On Logical Connectors (Junctors)

The logical significance of connector or "syncategorema" made visible in the pyramid – Deficiencies of Wittgenstein's and Post's truth-value tables of the logical connectors or so-called logical constants – Proposition-forming connectors characterized by truth values as opposed to expression-forming connectors which lack truth values – How each of them is constructed in the pyramid – The distinct functions of each of the various logical connectors in the pyramid – Equivalence as an expression-forming connector which lacks any truth value. Its function in definitions – Mathematical equations as definitions lacking truth-values – The basic mathematical operators as logical connectors for the construction of arithmetical expressions – The multiplication operator as fusing, rather than connecting concepts – The logical functions of the other mathematical connectors in defining and producing mathematical concepts or expressions: sum, difference, two types of quotient-construction – The logical significance of differential quotients and integrals – The modal connectors "probably" and "possibly" suspend the assertive character of propositions

2. Logical connectors are logical concepts for the description and interpretation of the relationships which obtain among the conceptual positions of the pyramid. As such, logical connectors have definite meanings and distinct scopes of application. The dominant opinion since Aristotle – that logical connectors are themselves "meaningless" and only express meanings in conjunction with expressions and propositions (Greek: synkategoremata; Latin, connotationes) – is therefore false. There are only vertical or horizontal relations. Logical connectors describe such relations in multiple ways. Relations of synonymy among some logical connectors are based on this multiplicity.
   2.1. Some logical connectors bind concepts together only into expressions which have no truth values. We name these expression-forming connectors. They have not been sufficiently reckoned with in previous logical investigations..
      2.1.1. The expression-forming logical connectors are the "and" (adjunctive connector), the exclusive and non-exclusive "or", the quantifying connectors and the equivalence connector. Negation is employed with

L. Geldsetzer and R.L. Schwartz, *Logical Thinking in the Pyramidal Schema of Concepts: The Logical and Mathematical Elements*, DOI 10.1007/978-94-007-5301-3_2,

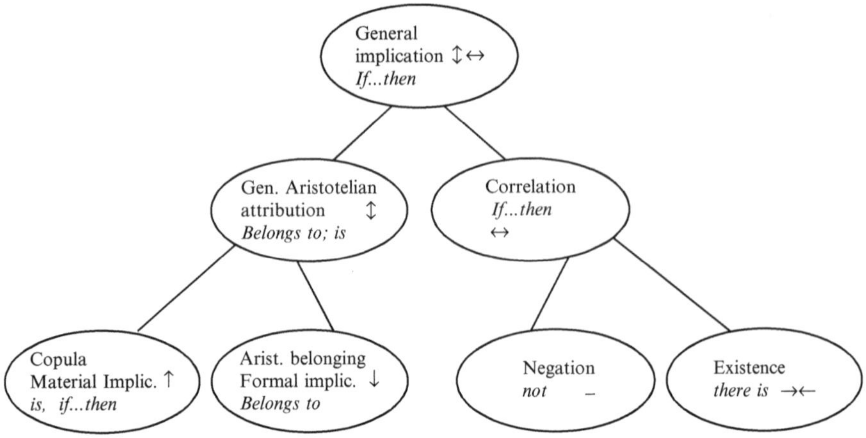

Fig. 2.1 Pyramid of the proposition forming connectors

so-called negative concepts as an expression-forming connector. Definite negation designates a (positive) concept by referring to its negated (dihaeretic) coordinate species (for example, "non-smoker").

2.2. The other logical connectors bind concepts together into propositions with truth values. We name them proposition-forming connectors. They are: The copula ("is"), the general and special (Aristotelian) attribution, the negation of the copula ("is not"), and the existential connector ("there is"). There are also four distinct implicative connectors (all alike formulated as "if..., then...."): 1. material, 2. formal, 3. correlative, 4. general implication. Among these implicative connectors the following are linguistically synonymous and therefore logically equivalent: Copula = material implication. Special Aristotelian attribution = formal implication or inclusion. The general implicative connector is equivalent to all of the other connectors taken together. That is why, in logical practice, the "if...then..." replaces connections which are sometimes true and sometimes false (Fig. 2.1).

As the conceptual pyramid which defines them shows, the proposition-forming connectors are themselves regular logical concepts. They connect regular concepts which are ordered within a pyramid. When they do so in the particular direction which is specific and proper (see the arrows) to each, the result is a true proposition; when the connection runs in any other direction, the result is a false proposition.

2.2.1. The truth-value tables of the connectors or so-called logical constants of L. Wittgenstein (Tractatus logico-philosophicus 5.101) and Post have become generally accepted standards of mathematical logic and especially of propositional logic. But they do not distinguish between expression-forming and proposition-forming connectors. They also attribute "meta-truth-values" for allegedly connected true and false elementary propositions to each of the 16 connectors constructed in the tables.

2.2.2. The truth-value table method of defining connectors involves several deficiencies. To begin with: "p" is supposed to signify "true proposition" and "not-p" is supposed to signify "false proposition". But "true proposition" and "false proposition" are not themselves true and false propositions; instead, they are specified concepts. Now, concepts have no truth values. The meta-truth values ascribed to their combinations are therefore not based on the truth values of (elementary) propositions, as is supposed by the method.

2.2.3. The meaning and application of almost half of the connectors defined in this way are dubious, and the proposals for naming and characterizing them are very unclear. There is no model or example for their meaningful application.

2.2.4. There is no definition of the copula, the proposition-forming connector ("is"). Instead the function of the copula is confused with the (expression-forming) function of equivalence. There is also no definition of the expression-forming connectors for quantification.

2.2.5. Tautology is "defined" as a joining together (adjunction) of two supposed truths ("if p then p, and if q then q"), but this amounts to the expression "one truth and another truth", which doesn't constitute any proposition at all. Contradiction is "defined" as an adjunction of two contradictions ("p and not-p, and q and not-q"), which means "a contradiction and another contradiction". Actually, these aren't definitions nor propositions at all.

2.2.6. Accordingly, the truth-value tables of propositional logic are incomplete, redundant, and misleading because they do not take account of the distinction between proposition-forming and expression-forming connectors.

2.3. The general implication is the highest level general concept among the proposition-forming logical connectors. Its meaning is the omnidirectional connection of concepts within a pyramid; that is, it is a connector which links concepts together in any direction. (If A then AB, if AB then A, if AB then AC, if AC then AB). It's truth values are those of the three subordinate forms of implication to the extent that these are true. It does not possess any truth value or falsity in its own right, but only to the extent it supplants one of the special implications.

2.3.1. The general implication is by no means tautological in the sense in which the tautology is defined in the truth-value tables as "always true" ("if p, then p" and "if q, then q"). Rather, the tautology cannot properly be considered a logical connector at all, simply because it doesn't connect. To "connect" something with itself is not a connection!

2.3.2. Unrestricted use of general implication would result in the apparent truth of any proposition in the form of an "if.., then..." statement. But such a use would contradict one of the false truth values of one of the forms of special implication. For that reason, such an unrestricted use could at most apply in cases where one lacks understanding of the thematic (pyramidal) connections of the concepts or propositions involved, as for example in the textbook piece of wisdom "If a butterfly beats its wings in Paris, than there will be a typhoon in Japan."

2.4. The two dihaeretic species concepts subordinate to the generic concept of general implication are the general Aristotelian "attribution," which connects vertically in both directions, and the correlating implication, which connects horizontally in both directions.

2.4.1. The general "attribution" is therefore ambiguous. It means that a generic concept or its characteristics as generic characteristics are intensional attributes of the species and subspecies which are subordinate to that generic concept. At the same time, it means that those species and subspecies are extensionally attributed to or subsumed under a common generic concept.

2.4.2. The correlative implication can be read as "if one, than the other." (If AB, then AC, and the reverse.) In substantive applications it serves as the logical formulation of correlative and in particular of causal relationships. In some Stoic inferences, it represents (as a conjecture which is not an assertion) the "non-demonstrable" premise for the causal inference.

2.4.3. The correlative implication presupposes a generic concept which is common to the correlated concepts. The common generic characteristics of the correlated concepts – in other words, that which is identical in each of them, also often called the tertium comparationis – are contained in this general concept, while the specific differences of the correlated concepts express that which distinguishes them.

2.4.4. When the correlative implication is applied to causal connections ("if cause, then effect" or "if effect, then cause") the temporal difference (cause before, effect after) of the linked substantive parts of the causal correlation must also be expressed in these specific differences. As Sextus Empiricus (flourished 180–200 A.D.) emphasized, cause and effect cannot be "simultaneous". In that case, the common superior concept designates a substantial identity between cause and effect which persists through time. It falls to the particular ontological theory in which this logical connector is used to specify in what this identity could consist.

2.4.5. The dihaeretic species concepts under the general attribution are: special attribution, which connects only from above to below (this is the connector which Aristotle employed most); and the copula, which connects only from below to above. The copula is synonymous with material implication (of generic characteristics in subordinate concepts).

Special attribution is synonymous with formal (extensional) implication or inclusion (of subordinate concepts in the extension of a superior one).

2.4.6. The dihaeretic species concepts combined in the concept of correlative implication are: negation (of the copula), on the one hand, and the existential connector, which connects horizontally in both directions, on the other. Negation occurs only between coordinate concepts (AB is not AC). Its image in the pyramid is the empty space between coordinate concepts. The existential connector, as negation's counterpart, cancels the distinction between coordinate concepts and converts their common (generic) characteristics into a higher level concept ("there is A"). That is why the existential connector serves to introduce general concepts into the logical formalism.

2.5. The negation (of the copula) is the distinguishing connector of coordinate concepts.

2.5.1. The negation of the copula is only reversible (commutative) between dihaeretic (mutually exclusive) coordinate species of a conceptual pyramid. It is not reversible between multiple coordinate species concepts, or between concepts drawn from different conceptual pyramids. In such cases it is transitive. Many logical mistakes of so-called double negations are due to the failure to observe this distinction.

2.5.2. The negation of the negation (so-called double negation) refers back to the initial concept only between dihaeretic species concepts. In all other applications double negation refers indeterminately to any and all concepts outside the scope of the general concept which includes the initially negated concept. The so-called "four corners" (Sanskrit: "catuskoti") of the Indian-Buddhist logician Nagarjuna (Second century A.D.) ("not A" and "not not-A" and "not: A and not-A" and "not: A or not-A") in their traditional understanding make use of this indeterminacy for the "proof" of the so-called "void" or "emptiness" (shunyata) of concepts.

2.6. The existential connector ("there is") designates directly individual conceptual positions within a pyramid or introduces their designations. When applied to content-filled examples, it asserts the existence of an object which corresponds to the concept. It also introduces negative expressions (negative terms) ("There are non-smokers"). Assertions concerning existence can be false if the existence of an existing (formalized) conceptual position is denied or a non-existent position is asserted. The conceptual position in the pyramid is itself the object of the logical assertion that it exists in relation to all other concepts.

2.6.1. The existential connector is often combined with negation, most often in the form of the negation of the extension of a concept. ("There are no ....") This mostly occurs, as a matter of both linguistic formulation and of content, within limited contexts. ("There are no fresh vegetables in winter," although of course there are in other seasons!). Absent such restrictions, the problem called "Meinong's paradox" arises, to the extent that the existence and non-existence of a concept (and in some cases of the object to which it refers) is asserted at the same time. In logic, however, this occurs only in the case of contradictory concepts. What is actually negated in the case of the negative existential connector is only the extension of the contradictory concept. What the negated existential connector expresses is that the contradictory concept has no extension of its own, that is, separate and apart from the combined extensions of its constituent concepts, and the same holds for all irregular (contradictory and dispositional) concepts. These concepts have traditionally been thought of as having no extension of their own and consequently referring to no object or state of affairs. Of course, a contradictory concept must be recognized as such in order to be introduced by means of a negated existential connector. For example: "There are no zombies"; that is, no living dead people.

2.7. Each connector which forms an expression is constructed by fusing together two of the dihaeretic connectors which form propositions, and which were discussed above, into contradictory connecting concepts (of the connector). For that reason, they possess no extensions of their own, but have at the same time both extensions of each of the original connectors which were combined to form them. Through this fusion they reciprocally cancel their respective functions, which generated truth values. That is why expression-forming connectors have no truth values. They can be inscribed, in rectangles, in the pyramid of the proposition-forming connectors as contradictory concepts. The expression-forming logical connectors also supply the logical basis of the mathematical connectors used in calculating (Fig. 2.2).

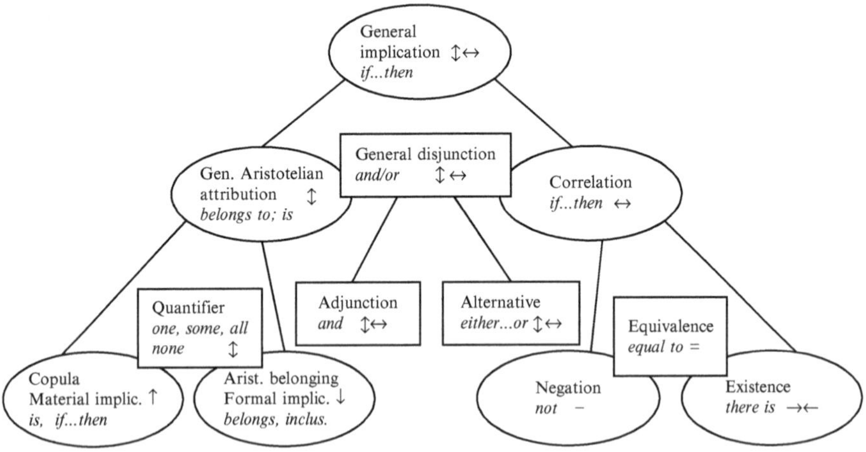

Fig. 2.2 Pyramid of both the proposition-forming and the expression-forming connectors

2.8. The non-exclusive (general) disjunction ("and/or" or Latin "vel") is formed by fusing general attribution and correlative implication and connects in all directions. Adjunction ("and") and the alternative ("either...or...", Latin "aut... aut") differentiate and specify these modes of connection. One sees that adjunction and the alternative are coordinate species concepts. They define each other reciprocally through negation: "And" means "not or" and "or" means "not and."

2.9. The (positive) quantifying expressions (a ..., some ..., all ...) are formed by fusing the copula and special attribution and they connect vertically in both directions. "all" connects a general concept with each of the subordinate concepts which come within its scope of application. "some" connects a general concept in an indeterminate form with one or more of its subordinate concepts, but does not express with which of two or more concepts the connection is made. "a" (one of) connects a general concept in an indeterminate fashion with one of its subordinate concepts or with one of the individuals which lie within its scope of application, but does not express, again, with which concept or individual of "all" of them, the connection obtains. Particular and individually quantified concepts are for that reason incompletely specified and require for their precise definition the exhibition of their intensions (through equivalence or in the form of an equation).

2.9.1. The logical "none" (negative quantifier) is usually formed by combining the negation and "a" or "one" ("not one") and is used in logic in that sense. But more precisely, what is meant is "not even one of" or "all not." "None" occurs frequently in the propositions of Aristotle's syllogisms (For example: "No animal is a plant"). But Aristotle's use can be understood throughout as the negation of a predicate ("All animals are non-plants"). In contrast, in mathematical logic the negative quantifier produces "empty concepts", that is, concepts which are said to have definite intensions but no extension. Such structures are in each case not logical concepts, but specifically mathematical ones, which are also imported into physics. That shows itself in the fact that they are, in contrast to logical concepts, subject to what is called Meinong's paradox: That which is thereby represented in consciousness doesn't exist, but if it did exist, it couldn't be represented!

2.10. Equivalence is often described as "reciprocal implication" ("then and only then when" or "if and only if"). As a logical matter, this usage can only refer to the generic characteristics of the correlative implication which is contained in, but constitutes only one aspect of the equivalence connector. In this context, to speak of implication is misleading to the extent that it suggests that equivalence involves two distinct coordinate concepts. That is not the case at all. For this implication exhibits, in addition to the generic characteristics of the (horizontal) implication, the specific characteristics of negation and the existential connector, each of which excludes the other.

Through this mutual exclusion their truth values neutralize each other, and it is this which the equivalence expresses. Negation signifies that both sides of the equivalence are not the same and therefore must be distinguishable. The existential connector, however, introduces both of the distinguished sides into the logical context of the formalism as "one concept." This equivalence is precisely what occurs when an incompletely specified concept is defined by means of one which is precisely specified (and the reverse).

2.10.1 Equivalence is therefore a connector which forms expressions, and that is why it cannot have truth values.

2.10.2. Since Leibniz it has been customary to consider all propositions with the same truth values as equivalent. Leibniz believed that all true propositions – simply because they were true – could be substituted for one another within a logical formalism without affecting either the validity of the inferences which contained them or, consequently, the truth values of the conclusions drawn from those inferences. (Leibniz's Latin formulation was "salva veritate" – "under preservation of truth"). But one doesn't see in what this equivalence between propositions which are true (and, accordingly, also between false propositions) – but which are also distinguishable and therefore distinct – could consist. The fact that they possess the common characteristic of being true (or false) does not render them any more equivalent than does the possession of any other common characteristic. Leibniz's view and its widespread application in logical formalisms is also false and misleading because "true proposition" or "false proposition" are, within the formalism itself, not propositions but specified concepts, which themselves can have no truth value. Genuine equivalence between propositions with the same truth values results only when they express the same propositional meaning (For example: "All animals are living things" = "Being alive is an attribute of all animals").

2.10.3. Equivalence expresses synonymy between concepts and terms. Equivalences are logical expressions without truth values. They serve as definitions, which can be "arbitrarily posited".

2.10.4. Lexica and dictionaries use equivalent expressions to define concepts. If equivalences were true propositions, all entries in (good) lexica and dictionaries would also be true. Anyone who wanted to rely on the exemplary mathematical "truth" of "two times two = four" could rely, with the same justification, on the exemplary philological "truth" that "poverty = (German) Armut."

2.10.5. Negated equivalences, for example those in etymological dictionaries (in which words with a current meaning are assigned to words with different historical meanings) similarly cannot have truth values.

2.11. Apart from multiplication (the formation of a product), the basic mathematical operations used for calculating are, from a logical point of view, logical connectors which form expressions. They are partly synonymous with logical connectors, and partly reducible to them. Multiplication is not a logical connector at all, but a fusion of concepts.

2.12. The principal mathematical connector is the equation. It is uncontroversial that equations are equivalences between that which stands to the left and to the right of the equals sign. Consequently, as a logical matter, an equation can have no truth value. Nevertheless, in mathematics equations are considered true propositions. Compare, for instance, Kant's example of a "synthetic judgment a priori": $5+7=12$, and Frege's example of an "analytically true function": $5+7=8+4$. Since elementary arithmetic equivalences have been, for centuries, memorized in the form of addition and multiplication tables and beyond that the expression "is equal to" (or simply "is") has usually been confused or identified with the copula, such elementary equivalences have come, as a matter of linguistic usage, to seem obviously true.

2.12.1. As a logical matter, inequalities are negated equations. Consequently, they too have no truth values.

2.12.2. The specifically mathematical relation "less than/equal to/greater than" ($< = >$) similarly has no truth value. Logically, it is synonymous with the nonexclusive disjunctive connector ("…or…or…or…" or "and/or….and/or….and/or…."). It unites an equation with inequalities in a single expression.

2.12.3. If equations and inequalities were true or false assertions, there would have to be also equations and inequalities in mathematics which would be both true and false propositions at the same time (that is, contradictions). But that does not exclude that mathematical equations contain sometimes contradictory expressions, as is the case in the formalization of geometrical functions as equations which are in fact correlative implications (see also 2.16.1).

2.13. Calculations in the form of equations define one or multiple numerical values (the "solutions" of the equation) and, in reverse, numbers define expressions used in calculations. The elementary calculations and multiplication tables learned in school are essentially memorization of some of these definitions and of the ways in which they are formed.

2.14. Beyond that, certain expressions for calculations define specific kinds of number concepts. For example, subtraction beyond zero results in the negative numbers, repeated division yielding continually recurring remainders results in the irrational numbers, finding roots of negative numbers results in the imaginary numbers.

2.14.1. Addition is a form of combination which is restricted to concepts which are coordinate species, that is, concepts at the same level of the pyramid. Subtraction is (in the positive range) a combination with a negated element ("and not"). Products are not formed by means of connectors but are instead specified quantities (for example, $3 \cdot 2 =$ "a doubled three or "a tripled two").

2.15. The other kinds of calculations define concepts or expressions as extended number concepts and metric concepts (particularly in geometry and physics).

2.15.1. Subtractions resulting in zero form the "empty concept" ("x and not x") and define it as zero (for example, 3–3=0). In analytical geometry, in which geometrical relationships (for example, curves in three-dimensional space) are defined arithmetically, each coordinate on the x, y-, or z-axis represents a particular continuum of positive and negative numbers with a zero point. The three dimensions of space intersect at right angles to each other in a common zero point. At the same time, however, each axis represents a continuum of zero points with respect to the coordinates located on the other two axes. This method of representing numbers is commonly called Cartesian system. But Descartes did not mention nor made use of it in his work "La géometrie" of 1637. Through this system zero is defined as the "common zero point" of the three number-continua as well as infinitely many zero points on each of the axes with respect to the other two axes. By this means, zero also becomes "definable" in functional equations by means of positive and negative numbers, which sounds quite dialectical. In a plane, for example, the points at which a curve intersects the x-axis (so that y=0) are included among the "solutions" of the functional equation y=f(x) of the curve. Applying this to the physical measurement of quantities permits one to equate quantities with zero value with positive or negative measured quantities; that is, to define the latter as equivalent to the former, which also sounds quite dialectical. However, this shows that the so-called geometrical functional equations, although formalized as and named equations, aren't really equations at all. Instead, they are mathematical propositions which correlate different numerical values of different Cartesian dimensions. This is the reason why the so-called solutions of such geometrical functions have truth-values. They are therefore to be formalized and read as implicative propositions, E.g.: "If y=0 then x=n".

2.15.2. Problems in subtraction which yield negative results form the "negative numbers" (in logical terms: numbers which are negated numbers, that is: not-numbers, but which nevertheless at the same time are numbers!). Their application to concepts of physics leads to negative concepts (minus degrees in temperature scales, for example, or positive and negative forces, positive and negative matter). The formation of such negative concepts has also become current in our "mathematicized" ordinary language. There is a plausible prototype for such usage in the language of banking and accounting. Debits and credits (liabilities and assets) are negative and positive concepts of financial value. The loan amount of the borrower, which for him is "in the red" is at the same time an asset or "in the black" for the bank, and the reverse is also true. In economics, one now speaks of "negative growth", which means a "positive decrease."

2.15.3. Mathematical powers represent the fusion of numerical concepts with themselves. This specific way of forming a product (by means of like factors) is used to define numerical values (which means other concepts). The procedure has been rendered plausible since antiquity through the model of geometrical squares and cubes with "sides of the same length." But this analogy fails to take account of the fact that these spatial "sides," considered as vectors, have different directions and are therefore different factors – a fact which stands in contradiction to the initial assumption of their identity. Moreover, any number raised to a power of one is defined as equal to itself, whereas geometrical square planes and cubes, the sides of which are equal to a single unit are plainly distinct from this single unit of length. In analytical geometry powers higher than the third power can be formulated whenever we wish to do so (in the so-called Minkowski spaces). But that which is to be represented as a fourth, fifth or higher power is only explained as a procedure analogous to combining a third Euclidian spatial dimension to the two dimensions of a plane. The physicist constructs concepts formed by means of powers "perceptible" only by means of models. Since he simply uses them as a formal factor in his calculations, he can, in principle, dispense with any sensory perception and hold out the "sensory imperceptibility" of such concepts as used in physics as their distinguishing feature. And that sets them apart from the concepts of other disciplines. Nevertheless, in other sciences and in philosophy there are concepts of the same type: That is to say, concepts formed by means of powers or "reflexive concepts," although their structure has scarcely been recognized. These concepts are characterized by the same dialectical doubling or self-multiplication, under simultaneous preservation of their unity. "Consciousness of self" (or self-consciousness) is a concept which satisfies these criteria. It has its prototype in the Aristotelian divine attribution of "thinking of thinking" (Greek: "noesis noeseos") and denotes both a unity and, at the same time, two separate things united with each other.

2.15.4. Quotients are ambiguous and have for that reason a double function. As proportions of quantities (some : some) they represent commutative expressions which cannot be further simplified by calculation (for example, the relationship between the number of goals scored by each team in a soccer match). As expressions for calculations in problems of division they define a part which results from a completed process of division. For that reason they are not commutative. (For example, distance/time = speed.). In the case of divisions which are "incomplete," that is, in which continued division of "the remainder" cannot be repeated with a finite succession of numerals, the quotient defines a specific kind of number, the irrational number. (For example, with respect to a circle, π (Pi) = circumference/diameter). The so-called "basic fractions" (1/2, 1/3, etc.) are themselves expressions used in ordinary language (a half, a third, etc.) and are therefore not problems requiring calculation. When they are represented in the form of equations (for example 1/2=0,5) what is shown is simply an equivalence between the representation of the expression as a quotient and as a decimal. In the case of 1/3=0,333…, where the number value of the right side of the equation is called irrational, one may benevolently speak of an equation but logically of non-equation, because nobody has ever shown what the exact decimal value really is. The mathematical term "limes" (Latin: boundary, German: Grenzwert) for the endless (infinite) series of points (which were never defined in mathematics as genuine mathematical signs) disguises this logical contradiction.

2.15.5. Differential quotients are proportions of infinitesimals, that is, "quantities" or "magnitudes" which cannot be expressed in numbers. Which sounds rather dialectical. Their meaning is customarily made clear – as it was by Leibniz when he invented them – through numerical proportions of the two smaller sides of a right-angled triangle, the hypotenuse of which as secant is fitted as closely as possible into a short segment of a curved line becoming in "infinitely short segments" a tangent. The differential quotient defines the so-called gradient of this curve at a point at which the shorter sides of the triangle and the hypotenuse are said to converge, approximating zero, in a way that can't be numerically determined. The contradictory nature of the differential quotient shows itself geometrically in the following way: Although this "point" is defined as "without extension," nevertheless, it is assumed that there are infinitesimal extensions at the same point (on the x- and y- axes of the Cartesian plane). This contradiction persists in the purely arithmetic application of the differential quotient, since there the expressions ∂x and ∂y serve as variables for "quantities" which are both numerical and infinitesimal. The usual explanation for this is that the infinitesimal "quantities" can approximate to zero without reaching it. This explanation camouflages the logical contradiction that such quantities are "simultaneously zero and different from zero."

2.15.6. Integrals add up certain sums – which are not capable of being added and which consist of numeric quantities and infinitesimal non-numbers – to form numbers. Since there isn't a method of calculation for the definition of integrals, their numerical values are defined by using sums of "exhausted" (that is, empirically derived) or of approximated quantities (of the content of planes or volumes) which are empirically measured and recorded in tables.

2.16. Functional equations (usually called simply "functions") are equivalences of expressions for calculations containing variables (so-called unknown magnitudes). The usual formal notation, y=f(x), expresses in logical terms that the general concept or numerical value y means the same as the higher-level concept x, to the extent that x is qualified by f as its specific difference. For example, "square = quadrangle with right angles and sides of the same length", or "bachelor = unmarried man", or "$4 = 2 \cdot 2$". That is the reason they are called equations. They were developed by Descartes as arithmetic definitions of geometric structures, specifically as definitions of strait or curved lines as rows of points in a plane. In Frege's terminology, one can formulate these relationships as follows: The sense of a quantified or specified value of x and the sense of a (likewise quantified or specified) value of y have the common reference of a point in a Cartesian plane. The replacement of the variables x and y by numerical values then defines a line as a sequence of points in the plane.

2.16.1. If one abstracts from these geometrical applications, functions lose, together with their geometrical meaning, their descriptive character. If x and y receive purely arithmetic meanings, which as a rule represent different numerical values, then the functional equations become correlative implications with truth values, that is assertive propositions. The art of "analysis" then consists in isolating the zero-value(s) of one of the variables and correlating the values of the other variable. Descartes, in his "Géometrie" recommended this procedure as "the best to consider" the equations.

2.17. Mathematical quantifiers which refer to units are synonymous with the logical "a" (in the sense of "one"). That applies as well to the so-called infinitesimal (infinitely small) which represents a non-numerical unit. Mathematical quantifiers which refer to definite numerical values (for example, "greater than one") are specifications of the logical "some." Mathematical quantifiers which refer to the infinite (large) are synonymous with the logical "all."

2.17.1. One sees from mathematical texts that in mathematics logical and mathematical connectors are used side by side. One moves freely in speaking of numbers and other mathematical entities from "one" to "some" or "all" or "none."

2.18. "Probably" or "possibly" are not logical connectors. But they do in fact play a significant role in the study of propositions and inferences. Their meaning and functions, however, have been mistaken, because it has generally been assumed, in both classical and mathematical logic, that even conjectures and predictions can only be expressed in the form of unequivocal assertions (since logic does not speak in the mode of the grammatical subjunctive!). When "probably" or "possibly" are added to logical connectors which form propositions, they suspend the assertive character of the proposition.

2.18.1. The use of the logical connectors which form implications also compels logicians to clothe conjectures in the form of true assertions. The result is that sentences properly expressed in the grammatical subjunctive (the counter-factual or the conditional) are formalized as propositions in the indicative, which convey an assertion. Logic would have avoided many mistakes and paths which have led it astray had it not followed Aristotle in logically formalizing linguistic connecting particles which, in ordinary language, express the subjunctive, such as “if … would have been (the case), then … would also have been (the case)” as implications expressing assertions, such as “if … is (the case), then … is (also the case)”. This was a mistake which the Stoics did not make.

Chapter 3
On Definitions

Definitions are equivalences and consequently expressions for synonymous meanings radically distinct from propositions with truth-values – The Aristotelian standard form of the definition and its deficiencies – Particularization as definition. – Algorithms as number concepts defined by particular number series and samples, and vice verse – Mathematical equations and geometrical (analytical) functions as definitions – Logical character of variables in equations – The example of Cartesian analytical geometry: three-dimensional number-representation in geometry – Dialectical character of so-called functional equations as definitions without truth-values and as implicative propositions with truth values

3. Definitions are expressions of equivalences. They are not propositions with truth values, as is generally assumed in logic. This false assumption confuses the connector which expresses equivalence with the copula. But the equivalence connector is properly read as the mathematical "equals." It was Jean LeRond d'Alembert who rightly observed this in his remarks on mathematical equations. It can also be expressed in ordinary language as "in other words" or "that means" or "respectively". What Frege said of equations is fundamentally true of definitions, namely that they express different senses of the same reference (meaning). Definitions explicate insufficiently specified concepts by using (more or less completely specified) "synonymous" concepts or expressions, and the reverse. Frege also confused the mathematical equation, which has no truth value, with the logical propositional assertion, which does.
   3.1. When definitions are formalized in logical terms, the formal signs on the left and right hand sides of the equivalence connector (that is, the equals sign) must be different. They represent or stand for synonymous concepts or expressions, which have an identical meaning (reference). Frege formulated this in a manner which was correct but hermeneutically deficient when he said that definitions express "different senses with the same reference".
      3.1.1. The reason that Frege's definition of the definition is incorrect from a hermeneutical standpoint is that Frege identifies a sign with its sense and distinguishes both of these from their meaning (reference). From the

L. Geldsetzer and R.L. Schwartz, *Logical Thinking in the Pyramidal Schema of Concepts: The Logical and Mathematical Elements*, DOI 10.1007/978-94-007-5301-3_3,

hermeneutical standpoint the correct formulation is that signs possess a double meaning. One of these meanings, (Frege's "sense") varies according to the formalism which is employed; for example, the sound value of letters, the numerical value of symbols for numbers, the meaning of connectors (including the connectors used as signs of operations in calculations), or the image represented by a graph. The other meaning is that which is properly termed logical or mathematical (Frege's reference). When the formalism is applied to actual states of affairs, the (linguistic or scientific) meaning of the concepts and expressions takes the place of this logical or mathematical meaning. One can also express this in customary hermeneutical terms by saying: Logical signs refer both to a surface literal meaning and to a deeper ("background") meaning established in a particular discipline.

3.1.2. If letters of the alphabet stand for entire concepts (as they do in the Aristotelian formalism) then A=B is a regular formulation of a (reciprocal) definition of synonymous concepts and only of such concepts. Different concepts which are dihairetic species define one another by negation: A=not-B. (For example: animal=non-plant.); B=not A (plant=non-animal.)

3.1.3. The expression "A=A" is neither an equation nor a definition. In textbooks, this formula is usually introduced (often in the formulation "A is A") as the first principle of logic, as an expression of the Principle of Identity and as a "tautology". In mathematics it is considered the prototype of the equation, so that all insertions of the same numerical value are held to be equations (3=3, for example). In such cases, however, the equivalence connector (and occasionally the copula) are being misused, in order to make signs or expressions, which do not define each other, appear to do so. In order to satisfy Frege's criterion for an equation the "A" to the left of the equals sign must have a different sense than the "A" to the right of it. But in the tautology both signs have the "same sense". Further, their common "reference" (meaning) must be distinguishable from their senses. In fact, however, sense and reference are the same in this tautology, and it has also been said to be a "name for truth". This mathematical use (or rather misuse) of the equation is one of the points at which dialectic enters into the notation of mathematical signs.

3.2. A standard Aristotelian definition circumscribes a species concept through the explicit statement of its generic intensions by means of the next-higher concept (genus proximum) and the "specific differences" (as pure intensions) which distinguish the defined concept from its genus and its coordinate species concepts. Such a definition describes the concept "distinctly" through its characteristics and demarcates it "clearly" from higher-level and coordinate concepts.

3.2.1. As a rule, the standard Aristotelian definition is incomplete. It declares the generic characteristics of a defined species concept to be only the specific differences of the next-higher generic concept (genus proximum) and takes no account of the characteristics which are generic for that next-higher concept itself. In order to be complete, a definition must state as well all generic characteristics of all higher-level concepts of a defined concept, right up to the most basic concept or category, which serves as an axiom.

3.3. So-called axiomatic basic concepts (highest-level genus concepts or categories) must be definable if they are to be concepts at all. Such highest genus concepts aren't definable according to standard Aristotelian definitions. However, since they contain the common characteristics of their species and subspecies concepts ("univoce" as J. Duns Scotus correctly saw, that is to say, as a uniform, unchanging meaning) they can and must be defined in another way.

3.3.1. When a word is available to express it, the designation of the axiomatic concept itself declares what this single characteristic is. Compare, for example, Duns Scotus' definition of "being" as the common characteristic of all of the (Aristotelian) "categories" which lie within the extension of the concept of being.

3.4. If such axiomatic concepts are contradictory, they contain only the characteristics of two dihaeretic general concepts, each of which excludes the other. (Compare: "possibility = being-nothing = nothing-being").

3.4.1. Aristotle's dictum that being is, according to the categories, expressed differently (Greek: "to on legetai pollachos") was interpreted by Thomas Aquinas to contain the assertion that the characteristic of the concept of being was contained in the categories as identical and simultaneously as distinct. This so-called analogy theory of being (analogy = simultaneous identity and difference) defines the concept of being as a contradictory concept. Duns Scotus' interpretation and theory of being as "univocal" (that is, as possessing a single meaning) defines the concept of being as a regular axiomatic concept.

3.5. So-called particular propositions (that is, those formed with the explicit or implicit quantifier "some") are definitions, not propositions with truth values. They express the equivalence between, on the one hand, a concept which is not sufficiently extensionally specified, and, on the other hand, another concept which indicates a proper specified extension.

3.5.1. Such definitions state at least two meanings for the explanandum (concept to be defined), such as "some A", according to whether the concept to be defined stands in a complete disjunction or in a sequence of multiple species concepts. An example of a dihaeretic relationship is: Some living things = animals/some living things = plants. An example of same-level sequence of multiple species concepts is: Some animals = dogs/some animals = cats/some animals = parrots, etc.

3.5.2. The arithmetical counterparts of particular concepts so defined are the so-called algorithms. A numerical expression containing variables is defined by the particular and/or individual numerical instances in decimal series or samples deducible from it, and vice-versa. Examples of such algorithms are expounded in 1.17.2. and 1.17.5. for the definition of decimal series and samples of odd numbers.

3.5.3. One of several indications that it is false to consider particular definitions as propositional assertions is that the result of negating them is that the positive terms which are to be explicated merely receive a negative designation. (For example: Some living things = non-animals/some living things = non-plants.) For that reason, such negations do not produce false propositions, but rather concepts defined by means of negation.

3.5.4. Another indication is that such particular definitions are often used to define past and future states of affairs, that is, objects of historical research and of experimental protocols, and of predictions, which all have the common characteristic of "non-being". One says easily: "Some things exist no more" or "not yet" without reflecting what this actually means.

3.6. So-called singular propositions are also definitions without truth values. They express the equivalence between a concept and its demarcation by means of generic characteristics and a singular extension of that concept (man = a rational animal). They are also used to define historical and predicted as well as present states of affairs. For example, Socrates = an Athenian philosopher. To wish to assert that "Socrates *is* an Athenian philosopher" – as often occurs when language is used without logical precision and when a definition is confused with an assertion, which alone is capable of being true or false – would simply be a lie, since everyone knows that Socrates no longer exists (see also 4.2.1.).

3.7. Mathematical equations and geometrical functions are likewise definitions. They express equivalences between a number and its paraphrase through expression of calculations. In a Cartesian plane, geometrical equivalences in the form of functions define sequences of points as lines.

3.7.1. From a logical point of view, numbers are substantive meanings designated by means of numerals used as signs (and in some cases through other mathematical symbols). In that sense, numerals as signs for numbers designate states of affairs within the context of mathematics as a discipline rather than logical states of affairs. The use of letters in mathematics since the seventeenth century is initially to be understood as the application of the Aristotelian conceptual formalism to the arithmetical objects. The result was that A, as the sign of a concept could as a matter of formal logic represent a number or an object. The equation "A = 2" defined the following state of affairs: Logically,

A stands for the number 2 and formally represents it; in other words, A, as the sign of a concept, is to have the same meaning as the numeral 2 used as a mathematical sign.

3.7.2. The use of letters as variables expanded the range of mathematical objects. It didn't at all mean that mathematical structures had been generalized or rendered logical. The use of lower case letters as signs for writing these "letter-numbers" (for example, a,b,c; m,n,o; x,y,z) was in part a recognition of this fact. When "a" was used as a variable, it no longer signified a logical concept of a number or numbers, but an indefinite or definite set of numbers. The (Cartesian) variables x and y (and in some cases z) signify likewise a mathematical object, namely one or more "unknown number(s)", which designate points within the Cartesian geometrical coordinate system. The so-called constants were intended as signs in the form of letters for specified numbers, or for what were believed to be specified numbers. For example, Pi (π) for the relationship of a circle's circumference and its diameter, or ∞ for an infinitely large number. In all these extensions of the range of mathematical objects the letters used as signs remained signs for numbers within the discipline of mathematics. In order to properly manipulate them, it was always necessary and possible to define their meanings through equations.

3.7.3. The traditional quadrivial application of arithmetic to the quantification of geometrical and physical structures provided an additional layer of meaning for these variables and constants as mathematical signs. At the same time, they were used as signs of logical concepts for geometrical and physical entities and themselves quantified by means of numbers. These double meanings of variables and constants became another conduit for the extension of the mathematical dialectic to geometry and physics.

3.7.4. The tautology is the simplest form of a dialectical mathematical equation. Wittgenstein wrongly believed that the tautology was the essence of the logical. (Tractatus logico-philosophicus 6.1.). It was through the tautology that the paradoxical character of the so-called principle of identity became canonical in mathematics. This paradox can be expressed in several ways. If, in the equation $x=x$, x is understood as "nothing," as was D. Hilbert's reading, then x is both a sign in the formalism and not a sign at all – because it signifies nothing. According to Hilbert, x meant "something" only when it has acquired this meaning through application to a numerical value or a concept. If x is understood as "itself" (following the Scholastic material supposition of the sign) then, too, x cannot be a sign at all, because signs, by their nature, refer to something other than themselves.

3.7.5. In technical mathematical terms one speaks of the "solution" or the "result" of a calculation which takes the form of an equation. But given such a "solution" it should no longer be written as an equation, but rather as the meaning of the equation, and consequently as a simple numerical value, herein corresponding to a properly defined concept.

3.7.6. Geometrical functional equations cannot possess "solutions" in this sense. What are called solutions in such cases are not the meanings of the variables, but the meaning(s) of one of the included variables when the value of the others is fixed at zero. This in turn presupposes that the defined "solution" refers to the Cartesian geometrical and sensorily perceptible model of three axes within a three-dimensional Euclidean space. Such geometrical functions are therefore not equations but rather correlative implications.

3.7.7. Cartesian analytical geometry defines in a geometrical and sensorily perceptible model three distinct kinds of numbers (along the three axes which stand at right angles to each other) which possess a common numerical value only at their common point of intersection at zero. For that reason, x, y, and z should actually be understood as indices or specifications of these three distinct number kinds (now called vectors). Each point in such a Cartesian space can be defined through its coordinates, that is, for each axis, by stating the numerical value of the distance of the (vector) coordinates from their common zero point: $P = x,y,z$. In mathematical usage these numerical values can also be quantified (in logical terms: specified). For example, $P = \frac{1}{2}x, 3y, 2z$.

3.7.8. What is logical about geometrical functional equations is that, for example, for each point on a line in the plane a value for x is paired with a value for y, and the same in reverse. The function $x = y$ is a readily understood instance of this. But if functional equations are employed purely arithmetically, that is, without reference to geometrical structure, a dialectical state of affairs results: Namely, that such equations define very distinct numerical values belonging to the x and y number kinds as identical values of one and the same kind of number. In this context one must accept, for example, that among the so-called solutions of these equations a zero value for one kind of number must be assigned as a coordinate to a numerical value which is distinct from zero for another kind of number, and that zero must therefore be identified with the respective numerical value distinct from zero. This certainly deserves the term dialectical.

3.7.9. To avoid the dialectic of such functional equations one must formalize them as correlative propositions, which are assertions. For example: If (some specified) x, then (some specified) y, and the reverse.

3.7.10. If we wish to use a logical example to clarify the essential character of a geometrical functional equation with two variables, we would have to say: The meaning x of a book, taken as a whole, defines the meaning y of a translation of that book, also taken as a whole, and the reverse, because both possess the same objective meaningful content. In the context of an arithmetical functional "equation", where one of the variables is reduced to zero, a proposition would have to obtain: If a book has any meaning at all, then it is identical with another book which has no meaning at all.

Chapter 4
On Propositions

Propositional assertions as the store of knowledge, which may be true, false or both at once – Mathematical equations are definitions lacking truth-values – So-called particular and individual propositions are also definitions lacking truth-values – Only proposition-forming connectors can properly join concepts or expressions to form propositions – Propositional truth-values appear plainly in the pyramid: Truth and falsity as flowing from the regular as opposed to the irregular use or reading of the connectors within the pyramid – Contradictory propositions are simultaneously true and false – The logical tradition - that they are "logically or analytically or a priori" false - is itself false – In paradoxical propositions truth and falsity are fused and reciprocally dependent – Alternative and disjunctive propositions are likewise simultaneously true and false, not "logically or analytically or a priori" true, as the tradition holds – On assertions and probable propositions and their relation to truth and falsity – Logical probability is generally considered as a third truth value. In fact, it has the structure of a contradiction, combining truth and falsity – It combines true general knowledge with specific ignorance about instances – The logical structure of the mathematical probability – But its contradictory structure is neglected thereby: Even the "most probable" or the "least probable" will either occur or not – "Fuzzy logic" as a kind of probability-logic

4. Propositional assertions constitute our store of knowledge. They can be true, false, or both simultaneously. Consequently, there must be knowledge which is true, knowledge which is false, and knowledge which is both true and false at the same time.
   4.1. Propositional assertions result from the linkage of concepts or expressions by means of proposition-forming connectors to form propositions. In mathematics and mathematical logic equations generally are considered to be propositional assertions (Compare, once more and for example, Kant's example of a "synthetic judgment a priori": $5+7=12$). But that view confuses equivalences or definitions with propositional assertions. The only genuine mathematical form of assertion is the purely arithmetic (and therefore non-geometrical) functional equation in cases where the numerical values of the expressions on the left and right sides of the equal sign are different. In those cases it is a correlative implication and therefore should not be formalized as an equation.

L. Geldsetzer and R.L. Schwartz, *Logical Thinking in the Pyramidal Schema of Concepts: The Logical and Mathematical Elements*, DOI 10.1007/978-94-007-5301-3_4,

4.2. So-called particular propositions, which classical logic since Aristotle has treated as assertions which may be true or false, are in fact definitions. They join a concept which is extensionally unspecified (definiendum: some x) with at least two meanings which stand in a negative relation to each other. (For example: some living things = animals, but also: some living things = plants, that is, non-animals.) If definitions are considered assertions, then one must accept that pretended particular propositions are simultaneously both true and false. (See also 3.5.)

4.2.1. So-called individual propositions (definiendum: one x) are likewise definitions. They assign a single meaning to a concept which is extensionally not sufficiently specified and thereby distinguish that meaning from the other individuals which fall under that concept. (For example, A dog = Fox-terrier Sofie, that is, no other than this one.) If one treats this as an assertion (a dog is the fox-terrier Sofie), then one must accept what is implicitly also asserted: An individual other than Sofie the fox-terrier is not a dog. (See also 3.6.)

4.2.2. In mathematical logic it is customary to combine particular and individual concepts requiring definitions with the connector which signifies existence. (For example: There is at least one x, so that $x = f(y)$). This is meant to be understood in the sense that it simplifies the so-called subalternation and makes it more precise. But this linkage with the existential connector rests on the mathematical conception that equations are assertions with truth values.

4.3. Propositional assertions connect concepts located within a pyramidal structure with one another. Simple propositions link only two concepts, as subject and predicate concepts, by means of a proposition-forming connector. Complex propositions link concepts and/or conceptual expressions (which have themselves been formed from concepts by means of an expression-forming connector) with one another. When the degree of complexity of the predicate (or some times of the subject) expressions increases, the result is so-called second- and higher-order predicate logics, as well as so-called relational logic.

4.4. True propositional assertions link regular concepts together within a conceptual pyramid by means of proposition-forming logical connectors used in ways defined to yield truths.

4.4.1. Wittgenstein believed that "what specifically characterizes logical propositions is that one can recognize their truth solely from their logical form, and this fact comprises within itself the entire philosophy of logic" (Tractatus 6.113). For traditional logical formalisms that is plainly not the case, but at most a logical ideal. It certainly is the case for the pyramidal formalism here proposed.

4.5. False propositions link regular concepts together within a conceptual pyramid by means of proposition-forming logical connectors used in ways not defined to yield truth. (E. g: "AB is AC" or "AB is not A" or "all A are AB").

4.6. Propositions which are both true and false (contradictory) link together regular (non-contradictory) and/or irregular (contradictory) concepts by means of proposition-forming connectors in such a way that the linkages may be read as both true and false.

4.6.1. These propositions are often not recognized as contradictory, particularly if the irregular concept they contain is also not so recognized. When that is the case, there is a conventional agreement on one of the two possible readings of the proposition. For example: "Socrates is mortal" = "Socrates is dead-alive." It is true that Socrates is dead and therefore false that he is alive. But the usual textbook example "All men are mortal" is conventionally held to be true.

4.6.2. A copulative proposition in which a positive concept is joined in the predicate expression with its negation is, as a whole, both true and false. (For example: "All AB are A and not-A".) This is the prototype of the contradictory judgment.

4.6.3. In classical and mathematical logic, such contradictory propositions are considered false. The result is that the true propositions which they contain are dialectically declared false, and this has had significant consequences in the history of science. What is here termed false is, taken more precisely, not logically false, since it is simultaneously true and false; rather, its content is confusing. Neither is it "meaningless" (absurd), since the instances of such a proposition must be clearly understood in order to be considered contradictory in the first place.

4.6.4. Contradictory propositions expressed in paradoxical formulations usually evoke perplexity. What essentially characterizes them is that they make truth depend on falsity and vice versa. For example, according to Parmenides everything which can be thought at all is thought as an unmoved unity, and consequently sensory perceptions of different things and of movement or change are false or deceptive. For the empiricist, who treats the sensorily perceptible as fundamental, it is on the contrary true that the arrow's flight is perceived as the movement of many arrows at different locations, and it must therefore be false that there is only a single unmoved arrow. Zeno, the disciple of Parmenides, could therefore formulate the proposition "The flying arrow is at rest" in order to defend Parmenides' "true" doctrine and to refute the empiricists. And the empiricists could employ the same proposition as "visibly false" to defend their own position and refute Parmenides.

4.6.5. Since paradoxes connect contradictory propositions, skillfully formulated, and are as such simultaneously true and false, any attempt to "prove their falsity" or to "refute" them is a permanent "makework" program without prospect of success. To "dissolve" paradoxes through "analysis" can only succeed through a demonstration that they are simultaneously true and false.

4.6.6. An alternative proposition is as a rule a copulative proposition in which the actual alternative consists in a predicate expression. (For example, AB is either A or non-A.) Such an alternative is likewise both true and false. In classical and mathematical logic, it is nevertheless considered true when one of the propositional parts it contains is true and the other false. This, too, has had and continues to have serious consequences in the history of science, since the false part of the proposition is thereby declared – dialectically – true as well as false. The sense in which an alternative proposition in contentful applications can be called true is at most a candid confession of ignorance as to which branch of the alternative is true and which false. It is therefore the usual means in which questions are posed for investigation. Compare the medieval method of "quaestiones", traditionally named the "sic et non" (Latin: "yes and no") method. In order to reflect our current understanding of the meaning of the logical connectors this method should now more properly be termed "sic aut non" ("yes or no").

4.6.7. In propositional logic, a false alternative is defined as one in which both propositional parts are true or both are false. But such a definition expresses only that such a structure is "not an alternative" at all. For an alternative is always simultaneously both true and false and for that reason has as such no unique value either of truth or of falsity.

4.6.8. The "non-exclusive disjunction" (Latin: " … vel … vel … ") is also in the rule a copulative proposition with a complex predicate expression. Since the disjunctive connector has two meanings, namely the meaning of the alternative and of the adjunction at once, this is precisely expressed as "… is … and/or …" in the language of officialdom. Disjunctive propositions are, like the alternative, true and false at once. The adjunction in them may relate to the true or the false parts and only to one of these, but never to both at once. If one uses disjunctive propositions in contentful applications one usually doesn't know which part (or parts) of them is true. One knows only that one part must be true.

4.7. Propositions expressing probabilities are nearly dominant in modern science. In considering them, however, one must distinguish precisely between (simple) logical probability, on the one hand, and mathematically quantified probability on the other.

4.7.1. If propositions expressing logical probabilities (in predictions, for example) are treated as assertions (and not as conjectures, which do not express an assertion) then they are negations of simultaneously true and false alternatives consisting of conflicting positive and negative statements concerning the same state of affairs. In that case, the logical meaning which the negated alternative assumes is that of the adjunction of both of the truth values. And such a negated alternative signifies that the assertion of a probability is to be understood not as being "true or false" (as in a two-valued logic) but as having a "third truth-value", which is "simultaneously true and false". Should the positive part of the assertion expressing probability later prove true (so-called verification), the negative part of the statement was false. Should the negative part of the assertion turn out to be true (so-called falsification) then the positive part was false.

4.7.2. Since a logical statement expressing probability contains at the same time both the verifying and the falsifying propositions one who uses it can – independently of whether it is later verified or falsified – always be right, or always be wrong, depending on the point of view.

4.7.3. Propositions expressing probabilities are the modern form of the old "docta ignorantia" (Latin: "learned ignorance") of Socrates, Nicholas Cusanus, and Blaise Pascal. Such propositions combine definite knowledge (even when it is only supposed knowledge) about the structure or interdependence of events with absolute ignorance concerning the forthcoming of a singular conjectured, predicted or hoped-for event.

4.7.4. That is why no one considers anything to be probable when one knows what is and what is not the case.

4.7.5. The exemplary empirical instance of a proposition expressing logical probability is a prediction of the result of a coin toss. What one knows with certainty is: Either heads or tails will come up, but never both at once. One certainly doesn't anticipate that the tossed coin remains standing on its rim; and if so, it would not be called a real toss. What one doesn't know is which of the two sides of the coin will come up. But what the logical statement of a probability asserts is: Both, heads and tails, will come up, but each of them with probability of one half (=0.5 or 50%).

4.8. Mathematically quantified propositions expressing probabilities are logically disjunctive propositions. The amount of knowledge increases with the number of instances involved, while the ignorance with respect to the particular instance – the coming up or not coming up of an event – remains.

4.8.1. So-called greater or lesser degrees of probability create the impression that they approach more or less closely to the truth. In fact such degrees refer to the quantum of knowledge contained in the proposition which expresses the probability, not to a greater or lesser proximity to what one doesn't know. That is why even that which is most "highly probable", just as much as what is least probable, still either occurs (turns out to be true) or doesn't.

4.8.2. The mathematical numerical formalization of probabilities rests on the assumption that the "degree of probability" is a calculable quotient of the relation of one event to all possible events of the same class of events. In this way one calculates the probability of a coin-toss as 1/2 (logically speaking: the probable toss is "either a AB or a AC"). The roll of a die has the quotient 1/6 (the probable case is "AB or AC or AD or AE or AF or AG"). But such calculations fail to take into account and cannot explain that and why in each such quantified probability the particular event either occurs or doesn't.

4.8.3. As a rule, quantified propositions expressing probabilities are copulative propositions with a disjunctive predicate expression with multiple predicates, which cover multiple, same-level species concepts ("possible cases") of a common generic concept (see the case of a die). However, the *mathematical* quantified disjunction of the possible cases masks the *logical* alternative which characterizes every instance of the possible cases (see the case of the coin-toss). The alternative being contradictory (see 4.6.6.), this results in making every probable proposition contradictory. The contradiction consists between the parts of knowledge and ignorance.

4.8.4. In quantified propositions expressing probabilities, the quantum of knowledge which they contain consists: (1) In the traditional belief of logicians, that a disjunctive proposition is true, when at least one branch of the linked alternative is false; (2) In the knowledge that the single event will actually occur in only one branch of the linked alternatives, but then cannot occur in the others; and (3) In the empirically (stochastic) surveyed or estimated or "a priori" constructed quantification of the occurrence of single events ("frequency") of each kind of event (The classic example is the equal frequency of each of the six sides of a true die when it is cast). The ignorance is as to: (1) What kind of event the particular event will turn out to be; and (2) If the particular event (irrespective of what kind of event it may be) will occur or not.

4.8.5. Knowledge relating to Nos. 2 and 3 (of 4.8.4.) must be formalized as a quotient which cannot be calculated (that is, a proportion, like the relative numbers of goals scored in a soccer game). When one casts the die, one therefore has to note the ("greater") probability 1:5. This expresses that each (hoped for, predicted, or "true") event is matched with five other (undesired, but likewise predicted or "false") events.

4.8.6. The probability quotient can also be reversed (as can the relationship between the number of goals scored in a soccer match). So, for example, if one wishes to avoid a certain result when casting a die, but the other five numbers are considered desirable. Then the greater probability of a fortunate throw should be noted as 5:1 instead of 6:1. One says: It is five times more probable that one will throw one of the desired numbers than the one undesirable number.

4.8.7. In the mathematical theory of probability the quantified probability is usually formalized as the calculable quotient of the occurrence of a single event in relation to all of the possible types of events (and thus-not just in relation to other types of events). The probability of rolling any single number on a die is specified as 1:6 (or 1/6). But it can also be formalized as the (relative) quotient (see 4.8.6., the probability of rolling a single number on a die is specified as 1:5 or 1/5). This applies as well to the customary conversion of the calculation into percentage terms expressing probabilities.

4.8.8. These divergent methods of quantifying probabilities show that they are completely irrelevant to what is actually of interest in propositions expressing probabilities, namely the occurrence or non-occurrence of the particular event of interest. Proof of these quantifications has customarily been sought by repeating test cases (throwing dies, for example) often enough so that the distribution of frequencies approaches the limit value of probability. But this begs the question, since one can and must already know the distributions are true, in order to be able to quantify the probability at all.

4.8.9. Whoever wagers a very small against a very large amount certainly knows that he will gain more if he wins the wager than he will forfeit should he lose it. But he doesn't know (unless the wager is rigged) whether he will win or lose.

4.8.10. The methods of quantifying probabilities just discussed merely trace the periphery of knowledge around a point where nothing at all can be known, namely the occurrence or non-occurrence of a single event or the presence or absence of a single instance.

4.8.11. Probability is also applied to relations between a part ("sample") and wholes or totalities. Inferences from part to whole are usually taken as exemplary incomplete and therefore insecure inductions. However, they are not inductions at all (see 1.8.), but propositional inferences. In these cases the characteristics of what is called a part or sample must be known. What one does not know is whether the same characteristics are generic characteristics of the whole and therefore contained also as generic intensions in all extensions (as parts) of the whole. If that's not the case, they can only represent specific intensions of the part or sample and distinguish it from other parts of the whole and from the whole itself. The inference results in a probable assertion that they are and are not.

4.8.12. If a sample or part is supposed to be "representative" of the whole, than "representativeness" means the generic characteristics of the part. And this must be inductively established beforehand. All parts of the whole must then contain the same characteristics, and this leaves no room for probable assertions but only for errors in the exact knowledge of what these common characteristics of the sample and the whole are.

4.8.13. Mathematical fuzzy logic (developed by Lofti Zadeh) is another instance in which an attempt has been made to construct a logic of lack of knowledge on the basis of considerations of probabilities. One assigns probability values to propositions with (qualitative) predicate concepts, the extensions of which are unspecified ("fuzzy"). As a logical matter, such "fuzzy" concepts cannot be concepts at all. Fuzzy logic is therefore precisely what its name signifies.

Chapter 5
On Inferences

Basic or simple inferences are conceptual implications and have truth values – The three kinds of simple inferences as formal, material and correlative implications have different truth-values – Only the correlative implication is appropriate for the formalization of causal relations – The dialectical character of self-implication as tautology – Aristotelian syllogisms consist of three propositions – The basic function of the "middle concept" in the premises is to guarantee the contextual coherence of meaning underlying the argument – They exhibit three patterns of conceptual implication: the "ladder", the "split" and the "summit" – Hypothetical syllogisms disguise non-assertive sentences into assertive propositional forms – Detrimental consequences of this custom in the traditional maxims: "Veritas sequitur ex quolibet" and "Ex falso sequitur quodlibet" – Stoic forms of argument (the Chrysippian "indemonstrables") are based on Stoic universal determinism – They combine a hypothetical "subjunctive" premise with various kinds of correlative implications between cause and effect, or vice versa – Propositional logic as mathematical calculus of inference – The Wittgensteinian truth-value tables define all connectors (including the expression-forming) by equations – Faults and errors about the expression "elementary sentence" and its use as variable in propositional logic – The "docta ignorantia" in probable inferences.

5. Inferences, which are currently often named arguments, are most often chains of propositions, like Aristotle's syllogisms and the Stoic (Chrysippus') inferences. Implicative propositions are the simplest form of inference.

5.1. These simplest forms of inference are of three kinds: (1) The so-called formal implication links a general concept with a subordinate concept which lies within the extension of the more general concept. It expresses the inclusion of lower concepts in the extensional range of higher ones. Its truth values are the same as those of the special Aristotelian attribution. (For example: "If living things, then animals" = "Life is attributed to animals"). (2) The so-called material implication links a concept with one of its higher-level concepts. It expresses the "material" identity of generic intensions in both higher- and lower-level concepts. Its truth values are the same as those of the copula. (For example: "If animals, then living things" = "animals are living things"). (3) Correlative implication may link any same-level concepts which fall under a common higher-level concept. Its truth values are the

L. Geldsetzer and R.L. Schwartz, *Logical Thinking in the Pyramidal Schema of Concepts: The Logical and Mathematical Elements*, DOI 10.1007/978-94-007-5301-3_5,

same as those of negation (for example: “If animals then plants” = “animals are not plants”; or “if plants, then animals” = “plants are not animals”) and of the logical connection signifying existence (“if animals then plants” = “there are living things, of which plants and animals are kinds”). Arithmetical functional “equations” with different values on the right and left sides of the equal sign are correlative implicative propositions despite the fact that they are not formalized as such but as equations.

5.1.1. Only the correlative implication is appropriate for the formalization of substantial causal propositions. Material and formal implication serve instead for the logical representation of logical justification or proof.

5.1.2. An example of non-discrimination between the logical functions and truth values of the various implications is Nelson Goodman’s paradox, also called “riddle of induction”.

5.1.3. In mathematics and mathematical logic, self-implication serves as the propositional formulation of identity (if x then x). Self-implication is considered always true. Wittgenstein held it to be the purest expression of logic’s tautological character. But if it is, the first x must be distinguished from the second x and therefore can’t be the same x. And at the same time the first x must be identical with the second x, and therefore indistinguishable from it. This is another example of the dialectical character of mathematical thought.

5.2. There exists a general implicative proposition which is exclusively true. It links concepts within a pyramid in any direction and thus contains in common the true linkages of the three special implications, without taking account of the instances in which each of them is false.

5.2.1. One should notice that, in the case of the implicative propositions as the simplest forms of inference, there can be no question of false or true parts of the inference, since such parts are merely concepts, not propositions (premises) capable of being true or false.

5.3. Aristotelian syllogisms are arguments employing three concepts; their intensional and extensional connections are clearly and distinctly represented in a conceptual pyramid through propositions and/or definitions. The so-called middle concept guarantees the thematic context which is explicated through the pyramidal connections. The relation of subsumption in the connection between generic, species and individual concepts used in the syllogism are represented by means of the copula or material implication (or both together) from below to above, or (as often by Aristotle himself) through special attribution or formal implication (or both together) from above to below. Their differences from neighboring concepts are represented by means of negative propositions.

5.3.1. Since in the Aristotelian syllogism everything depends on the subordination and differentiation of the three concepts, quantification plays the principal role. Differentiation of neighboring concepts can be represented by means of the negated copula (“is not”) or through

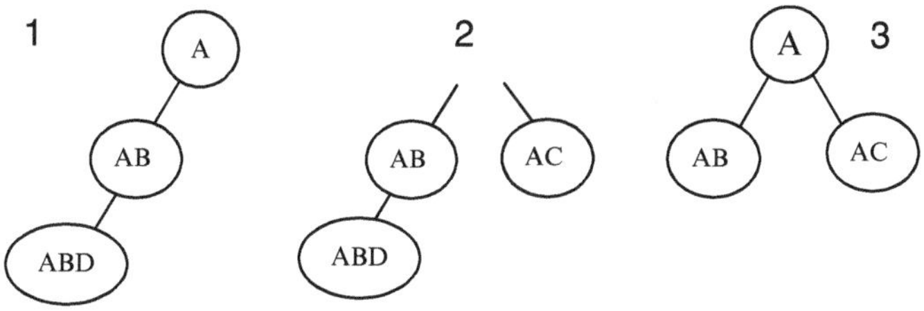

Fig. 5.1 The three schemata of Aristotelian syllogisms (1. ladder, 2. split, 3. summit, Example: A = living being; AB = animal; AC = plant; ABD = dog)

the negation of a concept ("no x is…"). The so-called particular or individual propositions which are contained in many syllogisms should be read as definitions (Instead of "is", read "in other words" or "is the same as" or "=").

5.3.2. All classical syllogisms can be reduced to and are constructed in accordance with three schemata. These are not the so-called syllogistic figures of Aristotle (Fig. 5.1).

5.3.3. In the "ladder", an individual or subspecies is linked to its species concept, the species concept with its higher-level generic concept, and in the conclusion the former is then linked to the latter.

5.3.4. In the "split", a subordinate species concept and a species concept are linked and both are distinguished through negation from a neighboring species concept.

5.3.5. In the "summit", two species concepts are distinguished from each other and both are linked to their generic concept.

5.4. Some or all of the sentences of a syllogism can also be expressed in the subjunctive. The distinction generally can't be recognized when the implicative connectors are used, but the connectors which are synonymous with them do make it clear. In the subjunctive cases, "if … then …" must be read as "In the case where …, … would be …". No assertion can be formulated by these means, only conjectures or hypotheses. And in such cases we are dealing with hypothetical or probable inferences, which Aristotle himself discussed. Like those Stoic inferences in the subjunctive which correspond to them, they are important heuristic instruments to be used in research. They serve to construct modal worlds.

5.4.1. In hypothetical syllogisms, the pyramidal formalism necessarily impresses the structure constituted by its logical connectors on the intensions which may be freely inserted in its conceptual positions. Customarily in logical practice, in order to preserve the functioning of the connectors with respect to truth values, this is distinguished as the so-called "validity" of the formalism from consideration of its (hypothetical) contents. One says that the logical formalism is valid, but the contents are false. For example: "If all men are birds, and if all birds fly, then all men fly." But to be precise one would have to say: "Were all men birds, and if all birds flew, then all men would fly." Then no one would misunderstand this conjecture as an assertion.

5.4.2. The traditional usage – to treat conjectures and hypotheses logically in the indicative as propositional assertions – leads to that odd definition of the truth of implicative inferences, which has had destructive consequences in the history of science: "An inference to a false conclusion from false premises is true", and "An inference to a true conclusion from false premises is true." Many strictly logical treatises, following that pattern, are then treated as "true" contributions to research on the basis of long chains of inferences. Similarly, the third definition of the implication, "An inference to a false conclusion from true premises is false" has often resulted in true and wide-ranging theories which serve as premises being discarded in their totality as "false" because of a false conclusion drawn from them.

5.4.3. Two traditional formulations of rules of implication – veritas sequitur ex quolibet (Latin: A true conclusion follows from what you will; that is, from either true or false premises) and ex falso sequitur quodlibet (Latin: Any conclusion, either true or false, follows from false premises) must therefore be used with caution. True conclusions follow only from premises – of whatever kind – which contain truth, although such premises may sometimes contain falsity as well. Arbitrary conclusions which are either true or false can follow from false premises only when such premises are understood as contradictions which by our definition contain both truth and falsity.

5.5. The so-called Stoic inferences (Chrysippus' five "indemonstrables") are arguments in which two concepts are either introduced within a pyramidal context by the existential connector ("there is …") or eliminated from the pyramidal context ("there is not …"). As shown above, the negation of the connector signifying existence refers formally only to the extension of the concept in question, substantially to the existence of a corresponding object or state of affairs.

5.5.1. According to the Stoics' view of logic, which was non-formal, the introduction or elimination of a concept in the pyramidal formalism corresponds to the substantial affirmation or negation of objects which are represented by the concepts in question.

5.5.2. Consequently, the Stoic inferences in a narrow sense assert the existence or non-existence of that which is represented by one concept as dependent on the existence or non-existence of objects which are represented by the other concept (so-called modus ponens and modus tollens) (Latin: the "positing" mode and the "denying" mode).

5.5.3. According to the Stoics' ontology, which presupposed a universal causal determinism, Stoic inferences can serve only for the logical clarification of relationships between causes and effects or effects and causes. But a cause-effect relationship can only be represented by means of a correlative implication between coordinate species concepts, since general concepts cannot be causes of lower-level concepts nor can lower-level concepts be causes of general concepts. The meaning of the five "indemonstrables" can only be explained with this presupposition. The premises of these arguments are plainly hypothetical conjectures which are formulated linguistically in the subjunctive. Otherwise, the conclusion would be just a repetition of the premise, and the Stoics expressly excluded that in the case of propositions and inferences. The following are the Stoic five "indemonstrables":

5.5.4. "If (in the event that) there were causes, there would also be effects. Now, there are (in the example in question) causes. Therefore, there are also effects."

5.5.5. "If (in the event that) there were causes, there would also be effects. Now, there are (in the example in question) no effects. Therefore there are (also) no causes."

5.5.6. "Causes and effects would not exist together at the same time. Now, there are (presently) causes. Therefore, there are (not yet) effects."

5.5.7. "There would be either causes or effects. There are (presently in the example in question) causes. Therefore, there are no effects (yet)."

5.5.8. "There would be either causes or effects. There are (presently, in the example in question) no effects. Therefore there are (only at present) causes".

5.6. If the premises of the Stoic inferences in the examples are read in the subjunctive, that is, as conjectures or hypotheses (If there were A, there would also be B), then the conclusion provides, as the result of an investigation, a verification or falsification of the conjecture ("Now there is A, therefore there is also B" or "Now there is no B, therefore there is also no A"). The Stoic forms of argument have therefore become, like the hypothetical syllogisms, heuristic tools for investigation and eristic patterns of argument. A numerous collection of these made in Indian antiquity can be found in the Madhyamaka-Karika of Nagarjuna (second century A.D.). In mathematics the corresponding (hypothetical) formula for introducing concepts and expressions is "Let ... be ..., then ...".

5.7. So-called propositional logic is usually considered to be the modern form of the theory of inference. Its inferences are said to consist in this, that from ("elementary") propositions which may be either true or false (and which are understood as assertions) the truth or falsity of an assertion which forms the conclusion is inferred. For example: "If p (true assertion) and not-q (false assertion), then not-r (false assertion)". Such inferences do not presuppose any thematic coherence among the individual propositions contained in the inference. In contrast, such a coherence is guaranteed both in the Aristotelian syllogism by means of the "middle concept", and in this book by means of the pyramidal structure of concepts. In mathematical logic these inferences are formulated as equations and such equations are mistakenly considered to be – as we showed earlier – propositional assertions. This view is false, on several logical grounds.

5.7.1. Equations are definitions. The whole of propositional logic (compare Wittgenstein's truth tables in the Tractatus) therefore provides only definitions of a supposed truth value of connected expressions. The only reason they appear plausible is that some of them have been selected in such a way that they agree with the truth values of the Stoic catalogue of proposition forms.

5.7.2. It is easy to understand why Wittgenstein never provided any example of an "elementary proposition." ("An elementary proposition consists of names. It is a connection, a linking together, of names." Tractatus 4.22.) This definition properly describes expressions which are connected with each other (such as: true proposition, false proposition) but which don't have truth values: it doesn't deal at all with propositional assertions, which have truth values.

5.7.3. One can recognize whether content-filled assertions are intended as true or as false statements (Assertions which are true and false, or contradictory, are treated in propositional logic as false!). When they are formalized by means of the signs for assertions or "elementary propositions" (normally "p" and "q") a truth value is attached to them (for example, p = true statement and not-p = false statement). But this kind of formalization confuses a negated proposition with a false one, despite the fact that negated propositions may be true. Moreover it converts the content-filled proposition – which is an assertion – into an expression, which can have no truth value. "True statement" is not itself a true statement, but rather a logical expression for a particular kind of statement. And the same is true of "false statement."

5.7.4. By formalizing implications of true and false content-filled propositions with the methods of propositional logic, such statements are not only converted into expressions without truth values. Beyond that, their thematic coherence, which was secured in Aristotelian syllogisms by means of the common middle concept, is lost, as is the logical distinction between premises and conclusion. We have become accustomed to examples of propositions used to argue in propositional logic somewhat like these: If Paris is the capital of France and three is a prime number, then January 13, 2011 was a Thursday (If p and q, then r). Or: If Paris is the capital of Morocco and 4 is a prime number, then January 13, 2011 was a Saturday (If not-p and not-q, then not-r). Both arguments are considered "valid" in propositional logic and consequently true. And that is declared to be so because a true conclusion follows two true premises in the first argument, and a false conclusion follows two false premises in the second. Clearly, in each such argument, the order of the statements is arbitrary, since none of them cohere thematically with the others and they therefore have no relevance as either premises or conclusions. In the case of content-filled propositions in propositional logic, it would be even more necessary to know – in advance of any supposed inference or conclusion – if the supposed premises and the conclusion were true or false, in order to determine the truth value of the argument. As a result, formalization with the methods of propositional logic is not only superfluous, but misleading.

5.7.5. Through formalization in the style of propositional logic all ostensible forms of statement, which are constructed with different logical connectors, become expressions which require logical definitions. For example, one of the four possible combinations of the so-called logical product is defined as "true proposition and false proposition = false proposition". Similarly one of the forms of the alternative is defined as "either a true proposition or a false one = a true proposition". One of the forms of implication is defined as "if a true proposition then a false proposition = false proposition" (more precisely a false or invalid inference). By means of these arbitrary (one should rather say: fantastic) definitions propositional logic has blinded itself for approximately 80 years to the plain fact that "true proposition" and "false proposition," in whatever combinations, are only logical expressions without truth values.

5.8. Probable inferences deduce an apparent knowledge from true premises in the logical form of a (contradictory) proposition expressing probability. Apparent knowledge means that in such inferences ignorance appears as knowledge, so that they are formalizations of a "docta ignorantia", a learned ignorance.

5.9. In mathematical logic numerically quantified probable inferences are recommended as instruments for the "rational" solution of decision problems under uncertainty as well as in competitive games and games of chance. Such problems present themselves in exemplary form in the relationship between diagnosis and prognosis of diseases, in business developments and political situations, and in lotteries, wagers and games of chance, and not least in legal procedure, where a "probability amounting to near-certainty" has often been a deadly trap for the accused. The theory of mathematical-statistical inference is intended to fulfil Comte's program of "savoir pour prevoir" (to know in order to predict). But here as well it is in fact a docta ignorantia (learned ignorance) in mathematical guise.

5.9.1. The clientele of the so-called logic of probability and of applied mathematics is misled when it is asserted that inferences concerning probability are instruments of valid (knowledge-expanding) inference and of the confirmation of hypotheses, or that in such inferences "certain probabilities are transferred from the premises (hypotheses) to the conclusion (thesis)", as we are told in dictionaries. All of the knowledge contained in such inferences consists in the premises, which are taken as true. The conclusion itself formulates the hypothesis or conjecture in the form of a contradictory judgment. Only afterwards does the ex post verification or falsification, which are themselves not hypotheses, confirm or refute one of the theses which are contained in the conclusion of the probable inference.

5.9.2. If probable inferences actually were – as their name suggests and as the majority view would have it – closer to truth than falsity, there would be no games of chance, gambling, lotteries, or stock market speculation.

5.9.3. Probable inferences are not suited to the discovery of unknown causal connections. No one knows if "everything" is subject to such causes. And if it were, as the Stoics with their universal causal determinism believed, and as classical mechanics assumed, then causality would be a generic characteristic of all species concepts and of each individual instance. That is why causality cannot itself be the object of probable inferences.

5.9.4. Neither can probable inferences demonstrate that "everything" is accidental or undetermined, as the Epicureans and quantum physics supposed. For such a condition would also have to be recognized by induction, as freedom or spontaneity, and would enter into all species concepts and all inductive frame concepts as one of their generic characteristics.

Chapter 6
On Theories

The conceptual pyramid is the "hard core" of theories – Defined uses of the connectors between concept-positions produce propositional assertions as statements of the theory – Competing theories of the same data derive from different inductions – On inductive and deductive theories – On contradictions in theories and on dialectical theories – On understanding and comparing different and especially competing theories – On true understanding of false theories and vice versa.

6. Theories are logically ordered conceptual pyramids covering a certain range of experience.

6.1. Insofar as a theory interests us only from the point of view of the concepts and definitions it contains, we are dealing with what J. D. Sneed and W. Stegmüller have highlighted as the "core" or a "non-statement view" of the theory.

6.2. The "hard core" of a theory, as it is formalized by the concept positions in a pyramid, can be read, by means of the logical connectors, as propositional assertions and inferences. This gives rise to the "statement view" of the theory.

6.3. In theories, (axiomatic) basic propositions are distinguished from theorems, which are deducible (demonstrable or provable) by means of the former. The model for this distinction is Euclid's book "Elements", which Euclid produced as a representation of the geometrical and arithmetical knowledge of his predecessors and contemporaries. The basic propositions are assertions, which contain the axiomatic basic concepts.

6.3.1. Theories can be constituted either inductively or deductively.

6.3.2. The concepts on which an inductive theory is based are lowest-level species concepts. In many theories so-called proper names and descriptions indicate only the specific differences of such lowest-level concepts. All other concepts constructed in the theory are to be inductively established or abstracted from these by leaving out specific differences and preserving common generic characteristics. This holds true particularly for the highest-level generic concepts (axiomatic concepts) of the theory. The "theory-ladenness" which

L. Geldsetzer and R.L. Schwartz, *Logical Thinking in the Pyramidal Schema of Concepts: The Logical and Mathematical Elements*, DOI 10.1007/978-94-007-5301-3_6,

K. R. Popper and others claim characterizes all empirical descriptions of facts and data rests, to the extent that it occurs, on the fact that such "theoretical terms" are themselves the products and results of common knowledge or previous scientific inductions, which are then available for use in deduction.

6.3.3. Idealism is since George Berkeley an inductive metaphysical theory concerning all of reality. To this reality belongs not only "being," which is defined inductively and by means of the senses, as Berkeley claimed, but also "nothing," which is also derived inductively from sensory experience.

6.3.4. On the one hand, deductive theories serve as a check on inductive theories by extending the regularities of the inductive procedure in the opposite direction. On the other hand, deductive theories can provide a formal framework for assigning terms, by means of which contents of knowledge which have not yet been formalized can be given a logical order.

6.3.5. In deductive theories, "dialectical" conceptual positions can be provided for. By this means, contradictory concepts can be defined and potentially fruitfully employed for heuristic use in the context of the core of the theory in question. This regularly occurs in mathematical theories; the pyramid of number concepts described above (1.16) is an example.

6.3.6. If contradictory concepts occur in deductive theories, their presence shows itself in the construction of propositions which contradict other propositions within the same theory. In mathematical theory, this regularly leads to paradoxes. Their discovery or deduction in mathematics is considered a signal achievement.

6.3.7. Realism has long been and still is a deductive and dialectical metaphysical theory. It contains the same assertions as the idealist theory, but opposes to each of its assertions a contradictory counter-assertion. For example, realism asserts that the being of things-in-themselves is only expressed and thought through consciousness. So far this accords with the idealist view, for idealism either does not distinguish between things-in-themselves and conscious appearances or it treats things-in-themselves as known (and insofar conscious) "noumena." But simultaneously, realism asserts that things-in-themselves are "independent" of all consciousness and ultimately unknowable. Realism is therefore not an alternative to idealism, but a dialectical expansion of it, adapted to the quadrivial natural sciences, which have always "consciously" sought to exclude "consciousness" as a subjective factor from supposedly objective knowledge, which must nevertheless presuppose it.

6.4. Contradictions in theories can be eliminated through re-examination and in some cases more precise definitions of the concepts involved. In the "logical criticism" of theories, it is customary but careless to consider an entire theory as false because a contradiction or paradox appears in it, and to reject it as a whole. Since contradictory propositions are simultaneously true and false, this results in the true propositions of the theory being discarded as well, something which has had significant consequences for the development of science.

6.4.1. Dialectical theories include, intentionally or unknowingly, contradictory concepts and propositions. Examples of the former are the theologies of Tertullian ("credo quia absurdum;" Latin: I believe because it is absurd, that is, contradictory); P. Abelard (Sic et non; Latin: Yes and no) and of Nicholas Cusanus (coincidentia oppositorum; Latin: the coincidence of opposities); Leibniz' "continuum law"; Hegel's dialectical concepts of human action; many of Nietzsche's coquettish aperçus (truth = error), as well as many Left-Hegelian theories and concepts (the "non-identical identical", "repressive tolerance," "peaceful struggle," etc.) In this book the existence and the essence of such contradictory concepts and propositions are taken seriously, from a logical point of view, and their logical relevance is analyzed. Examples of the latter (unintentional use of contradictions) include modal logic and the so-called many-valued logic as well as large parts of mathematical theories and, because of their use of mathematics, of physical theories as well. Occasional contradictions in the concepts used in legal doctrines, for example, "unwritten law", "penalty-free crimes", "legal personality"(= "non-personal personality") result from legislative compromises which make use of such formulations. They have never resulted in the doctrines themselves – or the theories which employs such concepts – being declared "false".

6.4.2. Examples for the pyramidal formulation of entire theories can be found in publications of Ralf Goeres and of L. Geldsetzer.

6.5. As a logical matter, theories have remained what the Greek word "theoria" originally meant, a "complete view" of a certain field of experience. The division of these fields of experience into disciplines has resulted in the "teaching" (doctrine) of the knowledge and insights collected in such disciplines being called a theory. A doctrinal theory consisted of concepts, propositions and inferences, and logic as a discipline taught what these are and how with them a "complete view" of the facts and data and their connections is to be articulated.

6.5.1. The philosophers of antiquity took first nature, then culture, and finally the entire world as their field of experience. In so doing they discovered that they brought forth completely different theories of their common world according to the different choices they made of inductive frameworks for their highest-level or most general concepts (Greek: "archai"). The differences among their "worldviews" became the model of all later pluralisms among theories in the disciplines and individual sciences.

6.5.2. Since Euclid's systematization of the geometrical and arithmetical knowledge of his predecessors and of contemporary research, mathematicians have held up as an ideal that there can be, in mathematics and in its later development into mathematical disciplines, in each field only a single theory, which is the true and correct one.

6.5.3. Physicists have adopted this ideal from mathematicians. Most recently, they have believed that there was no alternative to the theory of relativity and quantum theory in physics, which they held out as the true theories of macro- and microphysics. Were that the case, it could only mean that for approximately 100 years there had been no essential theoretical progress in physics.

6.6. Theories are not distinguishable on the basis that they are expressed in different natural languages. If a particular language lacks the vocabulary or means of expression which a theory requires, a genuine translation will require either their invention or the importation of relevant technical terms borrowed from other languages. By this means the various culturally developed languages enrich each other.

6.6.1. Theories which involve different "basic" or lowest-level concepts within the domain of a particular discipline distinguish themselves by that very fact as competing theories within the discipline. The exemplary case is where a certain phenomenon is defined as a (empirically observed) fact in one theory, and as an artifact in another. Generally, artifacts are eliminated from the domain of relevant facts by means of Ockham's razor.

6.6.2. But different theories can also be constructed concerning one and the same set of facts, as was shown above. (1.9.1. Fig. 1.3.) This can be observed equally well with respect to both contemporaneously competing theories and those developed in different historical periods, particularly with respect to the relationship of a predecessor to a current theory within a particular discipline. The differences among competing theories derive from different selections of concepts to serve as inductive frameworks. For example, Aristotelian-Ptolemaic astronomy defined the concept of the earth as a counterpart concept (a co-species) against the concept of the heavenly body. Copernican theory defines the earth as an individual of the species planet, which in turn falls under the genus of the heavenly bodies. It is the result of this different placement of the relevant theoretical concepts within a pyramidal scheme that the same statements about the earth and the heavenly bodies are assigned opposing truth values in the two different systems. What is read off from the conceptual pyramid and asserted in a statement is true under one of the theories and false under the other, and the same holds in reverse.

6.6.3. The falsity of a theory – or of some of its statements – as measured against what is considered a true theory is not an argument that such a theory is, as has been much discussed, incommensurate, incomprehensible, meaningless, or untranslatable. Putting aside instances in which theories contain contradictions which are simultaneously true and false, statements or theories which are simply false must be well understood in order to be recognized as such. They can therefore not be incommensurable, incomprehensible, meaningless, or untranslatable from the standpoint of another theory (which is generally considered to be true).

6.6.4. It suffices, in order to adequately understand false theoretical statements, that they are understood to not be true. Such an understanding indicates that the concepts which constitute the statements are located in pyramidal positions which differ from those which underlie true theoretical statements. If such statements were understood as being "neither true nor false", that would amount to excluding them from science altogether.

6.6.5. Within a particular scientific discipline, a bibliography does more than register the titles of relevant publications. It also supplies a "topic" – an overview of the relevant concepts of that science. These may be fairly termed the "hard core" of all of the theories of that science.

Chapter 7
On Axioms and Especially on the Real Axioms of Logic

Axioms or "categories" are not indefinable as the tradition holds – Implicit definition of axioms as petitio principii – The real meaning of the traditional logical "principles" of Identity, of Contradiction and of the Third – Critique of the traditional uses of the three "principles" in logic and mathematics – The "principle" of the Third is itself contradictory and therefore logically superfluous – On the non-applicability of the so-called criteria of axioms – Undecidability results from "indefinable" or dialectical axioms – Identity, contradiction and the third not being axioms, they are rather deduced concepts of logical theories – On truth, falsity and truth-falsity (or probability) as the genuine axioms of logic – Definition of truth as pyramidal coherence of concepts, falsity as pyramidal incoherence of concepts – Truth-falsity (or probability) is a contradictory fusion of coherence and incoherence – Literary fiction is "neither true nor false", that means "not true and not false" or "false and true" – Advise how to interpret it – Thought experiments as "science fiction" and how to interpret it – Logical truth is no substitute for scientific trustworthiness.

7. Axioms are highest-level general concepts of theories, which are constituted through the pyramid or pyramids of all of their lower-level concepts. That is why they must be formed inductively in order to be regular concepts at all. However, the Stoics believed them to be cognate foundations of thought common to all human beings (Greek: koinai ennoia; Latin: notiones communes, "common places of thought"). Euclid, following Plato, considered them innate ideas. However, according to the classical type of Aristotelian definition, highest-level general concepts cannot be defined. This (false) Aristotelian view turned out to be an invitation to proceed in the construction of axiomatic systems according to the maxim: "Adopt as an axiom whatever you can't explain!" At best, their so-called implicit definition consists in the demonstration that their intensions are contained in all of the lower-level concepts under them as generic characteristics. And that generally amounts to a "petitio principii" (begging the question), since in the process that which is deduced from the axiomatic basic concept has been inductively inserted into it.

7.1. The concepts identity, contradiction and the "excluded middle" or "third" (neither identity nor contradiction) are considered the highest-level general

L. Geldsetzer and R.L. Schwartz, *Logical Thinking in the Pyramidal Schema of Concepts: The Logical and Mathematical Elements*, DOI 10.1007/978-94-007-5301-3_7,

concepts of logic. However, the logical norms or rules which govern their application in logical relationships, namely to preserve identity, avoid contradiction, and exclude the third, do not demonstrate that they are in fact highest-level concepts. Non-compliance with these norms results either in other types of logic or in some cases in non-logical thought. That shows that identity, contradiction, and the third are not logical axioms, but species concepts derived within theory of logic.

7.1.1. What identity, contradiction and the third (apart from methodological maxims to preserve, avoid and exclude!) mean in logic can be defined as follows: Identity = the meaning of like signs in a formalism. Contradiction in a concept = the fusion of completely disjunctive (dihaeretic) concepts into a single (dialectical) concept. Contradiction in proposition = an adding together of a positive general proposition and its negation, which is both true and false. This can also be expressed in a general proposition where the predicate is an adjunction of a concept and its negation (e.g. AB is A and not-A). The third = negated alternative proposition (AB is not: A or not-A = contradictory adjunctive proposition AB is A and not-A).

7.1.2. Negation of an alternative proposition is synonymous or equivalent with the assertion of an adjunction (for example, "AB is A and not-A", see 7.1.9.). To that extent such a negation has the same meaning as a contradictory proposition. This shows that the third or the supposedly excluded "middle" as a logical truth value (as distinct from either truth or falsity) is itself contradictory and, to that extent, superfluous as a generic concept within logic.

7.1.3. In both classical and mathematical logic, identity, contradiction and the third are employed as forms of notation of the so-called truth values ("true", "false", and "third and other truth values" as distinct from either true or false). These are of course only applicable to propositions, since concepts as such have no truth values. It is usually said (and to be found in almost all logic textbooks) that: (A) A proposition expressing identity is "always true". The ground of this identity (which is no identity at all, see 3.1.3.) is explained as "purely formal", "purely logical", "tautological", or "a priori". (B) Contradictory judgments are "always false", and this (false prejudice) is explained in the same way. (C) The third or "middle" and other truth values are introduced and treated as "quantified probabilities".

7.1.4. As shown above, logical identity, with respect to any logical symbol, can be properly represented through repeated use of that symbol. But it is implication that is usually employed to express logical identity in propositional form (for example, if x then x, but see 3.1.3.). Contradictory propositions are, in contrast to the traditional opinion, not false; they are simultaneously true and false, or true-false (for example, "x is y and not-y"). Propositions expressing quantified probabilities are, as shown, contradictory propositions and, for that reason, simultaneously true and false (for example: "x is (with n % probability) y and (with 100 – n % probability) not-y").

7.1.5. The expression x = x is a (inadmissible, see 3.1.3.) definition of identity by means of an equation, and therefore not a proposition. Leibniz, in accord with his Principle of the Identity of Indiscernibles (called "Leibniz' Law"), defined identity as the indiscernibility of two "x's" to the left and right of the equal sign. But at the same time he distinguished the left x from the right x and thereby expressed their discernability. Consequently, this equation defines identity in contradictory fashion as "identity and non-identity". And it is this contradictory character alone which accounts for the fact that one can use mathematical systems of equations to derive or deduce both identities and non-identities from this contradictory concept of identity.

7.1.6. Accordingly, Frege deduced, for all equations, the "difference of sense" (non-identity) of x and y and the simultaneous "sameness of reference" (identity of meaning) of x and y.

7.1.7. The expression "x → x" (x follows from x, or x implies x, or if x then x) is intended, as distinct from the definitional equation x = x, to express an assertion about identity as a basic principle of logic. But what has already been said as to the definitional equation applies: The so-called principle of identity is necessarily, in this formulation, a contradictory proposition which simultaneously asserts the identity and difference (= non-identity) of x.

7.1.8. The expression "x = y and not-y" (x = y and ¬y) is considered the definition of a contradictory concept; and "x is (simultaneously) y and not-y" is accordingly considered the formulation of the principle of the (avoidance of) contradiction. In the context of logical axioms, this expression is thought to be the negation of the (supposed) principle of identity, x = x, and is therefore formulated as "x = not-x" or "x ≠ x". But since each logical symbol apart from x itself denotes an instance of not-x, it would follow that each equation of the form x = y would also be a contradictory expression, precisely because x ≠ y would be a non-contradictory expression of identity.

7.1.9. The expression "x = either y or not-y" ("x = y or ¬y") is generally thought to be the definition of the (excluded) third or middle. This expression is assumed to mean that there is "no third instance apart from (or middle between) y and not-y. Expressed with the copula, this is "x is either y or not-y," which figures as the propositional principle. But it can easily be seen that the formal expression only signifies an alternative, and that the assumption doesn't in any way define what the middle is. The correct definition of the middle which fits the subject matter consists in the negation of the alternative (x = not: either y or not-y). But this definition is synonymous with the adjunction "x = y and –y". And this shows that the middle – in the logical sense – signifies the contradiction itself. The middle is therefore not an independent principle, apart from identity and contradiction, and is consequently superfluous in logic.

7.2. The mathematical theory of axioms has proposed criteria for axioms. But these are neither themselves axioms nor can they be at all meaningfully applied. These well-known criteria are independence (from one another), freedom from contradictions, completeness, and possibly evident character. But if axioms are to be independent of each other, they cannot be checked for freedom from contradictions. If they are to be free from contradictions, they cannot be independent of each other. If they are to be complete, they must be the product of a complete induction, and all of the propositions of the resulting theory must be known as derived from these axioms, and justified. If they are to be evident (in any sense whatever), their intensions and extensions must at least be known.

7.2.1. That the "principia mathematica" or the "principles of mathematics" in B. Russell's and A. N. Whitehead's work of the same name (3 vols, 1910–1913. 2nd ed. 1927) contain theorems (that is: well-formed expressions) which are not deducible from their axioms and which are therefore indeterminate as to their truth or falsity, according to K. Goedel's diagnosis, results from the fact that these axioms consist of (unacknowledged) contradictory basic concepts and propositions formed with them from which both the truth and the falsity of mathematical theorems can be deduced. Undecidability then means that one can take at will a deduced proposition (or well formed formula) as true or as false.

7.3. The actual logical axioms are the concepts truth, falsity, and truth-falsity or probability (German: "Wahrscheinlichkeit"). One can name *truth* the first principle, *falsity* the second, and *truth-falsity* the third. The name "Wahrscheinlichkeit" (in German, this term as translated from the Latin "verisimilitas" contains both the word for truth and the word for appearance, so that it can literally be translated as "that which appears true") expresses the fusion of the concepts of "truth," on the one hand, and "false appearance," on the other, in our understanding of probability.

7.4. Truth and falsity are the highest-level regular generic concepts of logic and as such its conceptual axioms. All logical propositions and inferences are located within their extensions. They stand in a relation of negation to each other. That is why these axiomatic propositions can be formed from them: Truth is not falsity; and, falsity is not truth.

7.4.1. The sole characteristic of the concept of truth signifies logical coherence or consistency. This manifests itself in the use and reading of proposition-forming logical connectors in conformity with their definitions as applied to regular concepts.

7.4.2. The sole characteristic of falsity signifies deviance from the definite truth-conducive connection between regular concepts. This manifests itself in the use and reading of proposition-forming logical connectors in ways inconsistent with their definitions as applied to regular concepts.

7.5. Truth-falsity or probability is a contradictory concept which arises from the fusion of truth and falsity. Within its extension lie all true and false as also all contradictory or probable propositions and inferences, because contradictory concepts have no proper extensions of their own, but rather the combined extensions of the initial concepts truth and falsity. The characteristics of truth-falsity and/or probability are the simultaneous consistency with and deviation from a truth-conducive reading of proposition-forming connectors.

7.6. Concepts, expressions and all linguistic articulations which are not assertions, such as questions, conjectures, commands, norms, and rules are not true and not false, since they don't fall under the logical principles of truth and falsity. This also holds, as was shown, for all definitions and the (genuine) equivalences in mathematical equations. Since particular and individual propositions are definitions, they also lack truth values (see 4.2. and 4.2.1.).

7.7. Fictional literature, including poetry, is mostly considered to be neither true nor false, although it is an essential characteristic of such literature that it is articulated in the form of assertions. Plato considered poetry to be a "lie" (and therefore false), Aristotle considered it "true." That each of these views retains a following today requires a logical explanation.

7.7.1. The explanation lies in the fact that the logically connecting expression "neither true nor false" (= not true and not false) also serves to reciprocally exchange the dihaeretic (mutually exclusive) concepts true and false. For what is not true is considered false, and that which is not false, as true. Fictional assertions, which are said to be "neither true nor false" (= not true and not false) must therefore be simultaneously "false and true." Accordingly, Plato and Aristotle are both correct in their views of poetry, namely that it contains both truth and falsities.

7.7.2. Since fictional literature is generally not considered from the point of view of its probability, one can and must distinguish that part of it which is true from that part which is false in order to understand it at all.

7.8. A significant part of scientific publications consists in "science fiction." So-called thought experiments represent that part of this "science fiction" which is of logical interest.

7.8.1. Science fiction literature consisting of thought experiments might currently be found with particular frequency in unpublished applications for grants in aid of proposed research projects. According to their logical nature such thought experiments cannot be either verified or falsified. The principles according to which they are judged can therefore be neither logical nor, insofar as logic is essential to science, scientific.

7.9. However, that does not mean that in science itself moral principles of truthfulness, of resistance to deception and trickery, and of caution, when one is assured that something is probable, do not have a decisive role to play.

Corollaries and Annotations to the Paragraphs

0. That logic is the "organon" or "instrument" of all sciences and erudition was always felt – although not always taken seriously – in the history of occidental science. Aristotle, who assigned it that role, did not clearly say whether logic itself should be a science (episteme) or not, and, if not, whether it could be carried on as a practical art as distinct from a (scientific) technique, since in Aristotle's time and for long afterwards the same term was used for both. (Greek: "techne" = Latin: "ars"). Only the Stoics expressly denominated logic a science. Consequently, logic developed in different times and contexts into all of these. But Aristotle did clearly assign mathematics as a "second episteme" to the theoretical sciences and situated it between metaphysics (or ontology) and physics. This suggested that mathematics also required logical instruments for its constitution. That suggestion was neglected for ages, but then adopted by modern "mathematical logicism". – Perhaps because Aristotle's logic has always been well-known and intensively studied, his teachings concerning mathematics, in contrast to those of Plato and Euclid, have been underestimated by the historians of philosophy. Some references: Joseph Biancani, Aristotelis loca mathematica ex omnibus eius operibus collecta, Bologna 1615; A. Görland, Aristoteles und die Mathematik, (Diss.) Marburg 1899; J. L. Heiberg, "Mathematisches in Aristoteles", in: Abhandlungen zur Geschichte der mathematischen Wissenschaften 18, Leipzig 1904, p. 1–49; Th. Heath, Mathematics in Aristotle, Oxford 1949; H. G. Apostle, Aristotle's philosophy of mathematics, Chicago 1952; I. Mueller, Aristotle on geometrical objects, in: Archiv für Geschichte der Philosophie, 52, 1970, p. 150–71; J. Lear, Aristotle's philosophy of mathematics, in: Philosophical Revue 91, 1982, p. 161–192; J. Barnes, Aristotle's arithmetic, in: Revue de philosophie ancienne 3, 1985, p. 97–133; E. Hussey, Aristotle on mathematical objects, in: I. Mueller (ed.), Peri ton mathematon, Edmonton 1991. – See also M. Cantor, Vorlesungen über Geschichte der Mathematik, vol. I, 3. ed. Leipzig 1907, p. 251–256.

0.1. On the problem of formalism in logic and mathematics see: L. Brouwer, Intuitionistische Betrachtungen über den Formalismus, in: Sitzungsberichte

L. Geldsetzer and R.L. Schwartz, *Logical Thinking in the Pyramidal Schema of Concepts: The Logical and Mathematical Elements*, DOI 10.1007/978-94-007-5301-3,

der Preußischen Akademie der Wissenschaften 1928, p. 48–52; R. Carnap, Formalization of Logic. Studies in Semantics II, Cambridge, Mass. 1943, 2. ed. 1959; S. Krämer, Symbolische Maschinen. Die Idee der Formalisierung in geschichtlichem Abriß, Darmstadt 1988; T. Stoneham, Logical Form and Thought Content, in: Analysis 59, 1993, p. 183–185; L. Horsten, Platonistic Formalism, in: Erkenntnis 55, 2001, p. 173–194; G. Brun, Die richtige Formel. Philosophische Probleme der logischen Formalisierung, Frankfurt a. M.-London 2003.

0.1.1. Inaugurators of "ideal languages" were G. Dalgarno (1626–1687), Ars signorum vulgo character universalis et lingua philosophica, London 1661, and J. Wilkins (1614–1672), An Essay towards a Real Character and a Philosophical Language, 1668, with their versions of "Characteristica universalis". See L. Couturat and L. Léau, Histoire de la langue universelle, Paris 1903. – G. Frege, Begriffsschrift, eine der arithmetischen nachgebildete Formelsprache des reinen Denkens, Halle 1879, B. Russell and A. N. Whitehead, Principia Mathematica, Cambridge 1910–1913, and L. Wittgenstein, Tractatus logico-philosophicus (3.325), 1921, continued to promote this view of logic, and it is due to their influence that it has become almost universal in modern logic. See also J. Sinnreich, (ed.), Zur Philosophie der idealen Sprache. Texte von Quine, Tarski, Martin, Hempel und Carnap, München 1972.

0.1.2. See Th. Hobbes: Elementorum philosophiae sectio prima: De Corpore, London 1655, Engl. ed. 1656, Part I: Logic; G. W. Leibniz, Dissertatio de arte combinatoria, in qua ex arithmeticae fundamentis complicationum et transpositionum doctrina novis praeceptis exstruitur, Leipzig 1666, also Frankfurt 1690; G. W. Leibniz, Specimen calculi universalis, and: Specimen calculis universalis addenda, 1681, in: Philosophische Schriften, ed. by C. I. Gerhardt, vol. 7, p. 221–243; E. Bonnot de Condillac, Logique ou les premiers développements de l' art de penser, Paris 1792; G. Ploucquet, Methodus calculandi in logicis, praemissis commentatione de arte characteristica, Frankfurt-Leipzig 1763; J. H. Lambert, Sechs Versuche einer Zeichenkunst der Vernunftlehre, in : J. H. Lambert, Logische und philosophische Abhandlungen, ed. by J. Bernoulli 1782; G. Boole, The mathematical analysis of logic, being an essay toward a calculus of deductive reasoning, London-Cambridge 1847.

0.1.3. On graphical or diagrammatic formalisms see: Stephanus Chauvin(us), Lexikon Philosophicum, 2, ed. Leeuwarden 1713 (repr. in Instrumenta Philosophica Series Lexica II, ed. by L. Geldsetzer, Düsseldorf 1967), Art. 'Arbor Porphyriana', p. 53–54; E. Hammer and Sun-Joo Shin, Euler's Visual Logic, in: History and Philosophy of Logic 19, 1998, p. 1–229; C. von Pückler, Rhematische Graphen. Über Peirce's Theorien der diagrammatischen Nachbildung von Propositionen, in: Philosophia Scientiae 4, 2000, p. 67–131; J. Venn, On the Diagrammatic and Mechanical Representations of Propositions and Reasoning, 1880; Br. von Freytag-Löringhoff, Neues System der Logik. Symbolisch-symmetrische Rekonstruktion und operative Anwendung des aristotelischen Ansatzes, Hamburg 1985; J. Sowa, Knowledge Repraesentation. Logical, Philosophical, and Computational Foundations, Pacific Grove 2000; B. Ganter and R. Wille: Formale Begriffsanalyse. Mathematische Grundlagen, Berlin 1996, Engl. transl.: Formal Concept Analysis. Mathematical Foundations, Berlin 1999; J. Norman, After Euclid. Visual Reasoning and the Epistemology of Diagrams, Stanford 2006.

0.2. A. J. Swinburne, Picture Logic. An Attempt to Popularize the Science of Reasoning by the Combination of Humorous Pictures with Examples of Reasoning Taken from Daily Life, London 1881. – For an example of pyramidal construction of Chinese characters see L. Geldsetzer, Grundriß der pyramidalen Logik mit einer logischen Kritik der mathematischen Logik und Bibliographie der Logik. Internet HHU (Heinrich-Heine- University) Düsseldorf, 2000, Introduction.

0.3. This view is commonly clothed in talk of "purely syntactical or operational" handling of "signs or symbols", which – nota bene – can't be signs or symbols at all without a minimum of semantic content! See also G. Abel, Signe et signification. Reflexions sur un problème fondamental de la théorie des symboles, in: Philosophia Scientiae 2, 1997, p. 21–35.

0.5. It seems remarkable that logicians have struggled for a 100 years against any kind of psychologism but not at all – at least not yet – against linguisticisms. Take the example of Edmund Husserl. He began his career with psychological reflections about mathematics in his "Philosophie der Arithmetik", vol. I, 1891 (see: Edm. Husserl, Early Writings in the Philosophy of Logic and Mathematics, transl. and ed. by D. Willard, in: Husserliana vol. 5, The Hague 1993), and – after B. Russell's critique – continued his work as a declared anti-psychologist with his "Logische Untersuchungen", 2 vols, Halle 1900–1901, 2. ed. 1928, repr. 3 vols, Tübingen 1968, 6. and 7. ed. 1993, Engl. transl.: "Logical Investigations", 2 vols, London-New York 1970. But he did so without any rupture in his so-called phenomenological methods. See also his self-characterization in: W. Ziegenfuss, Philosophen-Lexikon, vol. I, Berlin 1949, art. "Husserl, Edmund", p. 569 – 576. – In contrast, linguisticism has had a lasting influence on logic after the so-called "linguistic turn". In this Wittgenstein took a decisive part through his Philosophische Untersuchungen / Philosophical Investigations, Oxford 1953, and later R. Montague through the so-called Montague-Grammar. See: R. Montague, Formal Philosophy. Selected Papers, ed. by R. Thomason, New Haven 1974. About this tendency see: G. Heyer, Eine linguistische Wende in der Logik? Bericht über den 7. Internationalen Kongreß für Logik, Methodologie und Wissenschaftstheorie 1983 in Salzburg, in: Zeitschrift für allgemeine Wissenschaftstheorie / Journal for General Philosophy of Science 15, 1984, p. 161–169. Talk of "logic as language" is still much *en vogue*, so it is high time to come back to a balanced relation between the use of language in logic (and other spheres) and the application of logic in linguistic matters.

0.6.3. H. Cappelen and E. LePore: Varieties of Quotation, in: Mind 106, 1997, p. 429–450; O. Müller, Zitierte Zeichenreihen. Eine Theorie des harmlos nichtextensionalen Gebrauchs von Anführungszeichen, in: Erkenntnis 44, 1996, p. 279–304; J. Pasniczek, On Bracketing Names and Quantifiers in First-order Logic, in: History and Philosophy of Logic 20, 1999, p. 239–250. – For the various scholastic conceptions of supposition see I. M. Bochenski, Formale Logik, 3. ed. Freiburg-München 1970, p. 186–199.

0.6.5. The oldest use of "meta-" with reference to relations among disciplines was by the editor of Aristotle's works, Andronikos of Rhodes, who placed Aristotle's writings on "Ontology" after or "beyond" those on "Physics" and called them "Metaphysics". And Meta-Physics was than mainly understood as a discipline treating the presuppositions of physics. D. Hilbert adopted the same use of the term with reference to mathematics in his famous lecture "Die logischen Grundlagen der Mathematik" (The logical foundations of mathematics) of 1923 where he conceived "eine gewissermaßen neue Mathematik, eine Metamathematik, die zur Sicherung jener notwendig ist, in der – im Gegensatz zu den rein formalen Schlußweisen der eigentlichen Mathematik – das inhaltliche Schließen zur Anwendung kommt, aber lediglich zum Nachweis der Widerspruchsfreiheit der Axiome" ("a to a certain extent new mathematics, a metamathematics required to safeguard mathematics itself, where – in contrast to the purely formal inferences of mathematics proper – substantive (i. e. contentful in distinction from purely formal) inference is used, but only for purposes of demonstrating that the mathematical axioms are free from contradictions"). See D. Hilbert, Die logischen Grundlagen der Mathematik, in: Mathematische Annalen 88, 1923, p. 151–165 (also in Gesammelte Abhandlungen vol. III, Berlin 1935, p. 178–191). In addition to Hilbert A. Tarski and J. Lukasiewicz proposed a "Metalogic" in: Untersuchungen über den Aussagenkalkül (Researches on the propositional calculus), in: Comptes rendues des séances de la Société des Sciences et Lettres de Varsovie, cl. III, 23, 1930 p. 30–50. See also H. Rasiowa and R. Sikorski: The Mathematics of Metamathematics, 3. ed. Warschaw 1970. – The dialectical character of this "meta-reflection" should be obvious. It was and is a "thinking about thinking", which Aristotle as "noesis noeseos" declared to be a divine capacity, and which now has become a typical feature of mathematical thought.

0.6.6. See A. Tarski, Der Wahrheitsbegriff in den formalisierten Sprachen, in: Studia philosophica 1, 1936, p. 261–405. Engl. transl.: The Concept of Truth in Formalized Languages, repr. in "Logic, Semantics, and Metamathematics", 2. ed. 1983

0.7. See Ch. P. Snow, Two Cultures and the Scientific Revolution (Rede Lecture of 1959), Cambridge 1963, new enlarged edition: The Two Cultures: A Second Look, with Introduction by St. Collini, Cambridge 1993. See also R. Wertheimer, How Mathematics isn't Logic, in: Ratio 12, 1999, p. 279–295.

0.8. G. Boole, The Mathematical Analysis of Logic, being an Essay toward a Calculus of Deductive Reasoning, Cambridge 1847, repr. Oxford 1948 and 1951, New York 1965; Aug. De Morgan, Formal Logic, or the Calculus of Inference, Necessary and Probable, London 1847; 2. ed. by A. E. Taylor 1926; G. Frege, Begriffsschrift, eine der arithmetischen nachgebildete Formelsprache des reinen Denkens, Halle 1879, repr. Darmstadt 1974, repr. ed. by I. Angelelli: Begriffsschrift und andere Aufsätze, 2. ed.. Hildesheim 1964, Engl. transl.: Conceptual Notation and Related Articles, ed. by T. W. Bynum, Oxford 1972; W. St. Jevons, On the mechanical performance of logical inference, 1870. – For a critique of these tendencies see G. Jacobi, Die Ansprüche der Logistiker auf die Logik und ihre Geschichtschreibung (The Claims of Logistics on Logic and its Historiography), Stuttgart 1962.

0.9. On the actual philosophical self-understanding of mathematicians see: M. Detlefsen, Philosophy of mathematics in the twentieth century, in: Philosophy of Science, Logic and Mathematics in the Twentieth Century, ed. by St. G. Shanker, London-New York 1996, p. 50–123; H. G. Carstens, Mathematiker und Philosophie der Mathematik, in: Zur Philosophie der mathematischen Erkenntnis, ed. by E. Bürger a. o., Würzburg 1981; M. Otte (ed.), Mathematiker über die Mathematik, Berlin 1974. See also D. D. Spalt, Vom Mythos der mathematischen Vernunft, Darmstadt 1981. – E. Agazzi (ed.), Modern Logic. A Survey. Historical, Philosophical, and Mathematical Aspects of Modern Logic and its Applications (Synthese Library 149), Dordrecht-Boston-London 1981, tried to give – on the basis of a "Rome Conference, September 1976" – a comprehensive survey of "modern logic" and its self-image. He tells us: "The purpose of the conference was to provide a general appreciation of modern logic which could be accessible to cultivated people, without need of any specific technical competence" (p. VIII). What the 25 contributions by renowned authors in fact show is that this "modern logic" has become and is understood purely as applied mathematics, so that "cultivated people" is tantamount to "mathematicians". – Against such claims we should recall Kant's recommendation in his Prolegomena: "Es ist aber eben nicht so was Unerhörtes, daß nach langer Bearbeitung einer Wissenschaft, wenn man wunder denkt, wie weit man schon darin gekommen sei, endlich sich jemand die Frage einfallen läßt: ob und wie überhaupt eine solche Wissenschaft möglich sei. Denn die menschliche Vernunft ist so baulustig, daß sie mehrmalen schon den Turm aufgeführt, hernach aber wieder abgetragen hat, um zu sehen, wie das Fundament desselben wohl beschaffen sein möchte". (It is not at all unheard of that after long cultivation of a discipline, when one thinks with astounishment how far one has developped it, someone finally raises the question, whether and how such a discipline is possible at all. For human reason is so prone to build that it has already many times erected a tower, only to demolish it afterwards in order to inspect the quality of its foundation). See I. Kant, Prolegomena zu einer jeden künftigen Metaphysik die als Wissenschaft wird auftreten können, ed. by K. Vorländer, Hamburg 1951, p. 2.

0.10.3. I have given an example of meaning-notation in Chinese characters in: Grundriß der pyramidalen Logik, Internet (Heinrich-Heine-University) HHU Düsseldorf, 2000, Introduction. That the same method of composition operates in the combination of Yin- and Yang-signs in the construction of the famous Pa Kua of the Yi Jing-Classic (Book of Changes) is shown in: L. Geldsetzer and Hong, Han-ding: Grundlagen der chinesischen Philosophie, Stuttgart 1998, chapter 4/2: "Die Logik des Yi Jing", p. 177–203. – On sensory foundations of logic see also: L. Eley, Metakritik der formalen Logik. Sinnliche Gewißheit als Horizont der Aussagenlogik und elementaren Prädikatenlogik, The Hague 1969, and G. Gutzmann, Logik als Erfahrungswissenschaft. Der Kalkülismus und Wege zu seiner Überwindung (Erfahrung und Denken 57), Berlin 1980.

1. R. Schwartz, Der Begriff des Begriffs in der philosophischen Lexikographie. Ein Beitrag zur Begriffsgeschichte (The Concept of the Concept in Philosophical Lexicography. A Contribution to the History of Concepts) (Phil. Diss. Düsseldorf), München 1983; also: E. Walter-Klaus, Inhalt und Umfang. Untersuchungen zur Geltung und zur Geschichte der Reziprozität von Extension und Intension, Hildesheim-Zürich-New York 1987. - Clearness and Distinctness have been since Descartes (see: Principia Philosophiae, Amsterdam 1644, I, § 45) and Leibniz (see: Meditationes de cognitione, veritate et ideis, 1684, in: Philosophische Schriften, ed. by C. I. Gerhardt, IV, p. 422) much debated characteristics of true sensory perception and / or metaphysical intuition. See: P. Markie: Clear and Distinct Perception and Metaphysical Certainty, in: Mind 88, 1979, p. 97–104. Both of them – and their followers – spoke about "ideas" (that is concepts), and took them to be true, if "clear and distinct". And both – and their followers – erred greatly in their supposition, that ideas or concepts as such could be true (or false). Due to their authority it never entered into the logical tradition that clarity and distinctness have nothing to do with truth or falsity, but rather with the extensional and intensional character of genuine concepts. A genuine logical concept must, by its extensions, be clearly distinguished from all other concepts¸ and its intensions must at the same time be distinctly grasped. This is what distinguishes our concept of the concept from what mathematical logic has understood to constitute a concept, following G. Frege, who maintained: "Unter Eigenschaften, die von einem Begriff ausgesagt werden, verstehe ich natürlich nicht die Merkmale, die den Begriff zusammensetzen. Diese sind Eigenschaften der Dinge, die unter den Begriff fallen, nicht des Begriffs" (I naturally don't understand the properties which are attributed to a concept to be the intensions which compose it. These are properties of the things which fall under the concept, not of the concept itself). See G. Frege, Grundlagen der Arithmetik. Eine logisch-mathematische

Untersuchung über den Begriff der Zahl, Breslau 1884, repr. 1934, p. 64. Engl. transl. by J. L. Austin, Oxford 1950.

1.1. See G. Berkeley, Treatise Concerning the Principles of Human Knowledge, Dublin 1710, especially § 97 and § 134. – The method of pyramidal formalization shows that concepts are formed by uniting distinct intensions with extensions. It also shows that propositions are actually formed in predicate logic by uniting pure intensions with those of a subject-concept having extensions. In contrast, the traditional conception of "whole concepts" was introduced and established in logic by Aristotle, who symbolized each concept with a single letter. This presupposed that propositional unity could only be achieved by joining a (whole) predicate concept to a subject concept. And this was the historical basis for the age-long "universals" debate over the ontological status of (whole) predicate concepts. – See D. M. Armstrong, Universals and Scientific Realism, Vol. 1: Nominalism and Realism; vol. 2: A Theory of Universals, Cambridge 1978; F. MacBride, Where are Particulars and Universals? in: Dialectica 52, 1998, p. 203–227; W. Stegmüller, Das Universalienproblem einst und jetzt, I and II, in: Archiv für Philosophie 6, 1956, p. 192–225 and 7, 1957, p. 45–81; W. Stegmüller, (ed.), Das Universalienproblem, Darmstadt 1978.

1.2. Induction was treated by Aristotle as (Greek) Epagogé, that is: concept construction progressing from particulars to the universal ("progressio a singulis ad universale"). He also compared this procedure with a syllogism, but remarked as well that it cannnot be a real syllogism because there is no "middle term" in it ("medio caret"). Besides, the epagogé is more convincing than any syllogism ("nobis manifestior est quam syllogismus"). Those are the traditional textbook doctrines of classical logic, and not at all mysterious. – Sextus Empiricus in his "Pyrrhonic Hypotyposeis", book 2, chapt. 15, criticised the "induction" in the following way: "Since (inductivists) want to warrant the general by individuals, they proceed either from all individuals or from some of them. But when they proceed from some of them the induction becomes uncertain, because of the possible neglect of some individuals which may not fit with the general. And when (they proceed) from all of them they attempt an impossible task, since the (numbers of) individuals are infinite and uncountable". This is the locus classicus for all later "sceptics" of induction. But Sextus went seriously wrong in supposing that quantifications of an induced concept should determine countable quantities of individuals (a view which Francis Bacon righly criticised, see 1.6.1.). The induced concept, by its intensions, determines once and for all which known or hitherto unknown individuals belong within or are excluded from its extension. And it is this meaning that mathematicians have adopted as "complete induction" (see also 1.8.1). D. Hume, J. St Mill and later inductivists, misled by Sextus' critique and Aristotle's suggestion, interpreted induction as a method of (syllogistic) inference on the basis of (causal) propositions. It was solely this propositional treatment of induction which was responsible for introducing the probability problem and other perplexities into the matter. But one should recall that in order to draw conclusions and construct inferences with truth values one requires beforehand concepts which cannot have any truth values. And they must be established by induction itself ! – See R. Lanton, Hume and the Problem of Induction, in: Philosophia (Philosophical Quarterly of Israel) 26, 1998, p. 105–117, and J. L: Mackie, Mill's Method of Induction, in: The Cement of the Universe, Oxford 1974. See further: N. Tsouyopoulos, Die induktive Methode und das Induktionsproblem in der griechischen Philosophie, in: Zeitschrift für allgemeine Wissenschaftstheorie / Journal for General Philosophy of Science 5, 1974, p. 94–122; W. Stegmüller, Das Problem der Induktion. Humes Herausforderung und moderne Antworten, in: H. Lenk (ed.): Neue Aspekte der Wissenschaftstheorie, Braunschweig 1971, p. 13–74 (together with: Der sogenannte Zirkel des Verstehens, Darmstadt 1975, repr. 1986 and 1991); H. Hoppe, Goodmans Schein-Rätsel. Über die Widersprüchlichkeit und Erfahrungswidrigkeit des sog. New Riddle of Induction, in: Zeitschrift für allgemeine Wissenschaftstheorie / Journal for General Philosophy of Science 6, 1975, p. 331–339; M. A. Changizi, and T. B. Barber: A Paradigm-based Solution to the Riddle of Induction, in: Studies in History and Philosophy of Science 30A, 1999, p. 419–484. – See also paragraphs 1.6. and 1.6.1.

1.2.1. See E. Cassirer, Substanzbegriff und Funktionsbegriff. Untersuchungen über die Grundlagen der Erkenntniskritik, Berlin 1910, 2. ed. 1923, repr. Darmstadt 1976, 7. ed. 1994. Engl. ed. Chicago-London 1923. – Thomas Aquinas, De Ente et Essentia III (Über das Sein und das Wesen, German-Latin ed. by R. Allers, Darmstadt 1965, p. 31): "quidquid est in specie est enim in genere ut non determinatum" (Whatever is in a specific concept is in an indeterminate way also in the general concept).

1.2.2. "Nothing" was defined by Aristotle as the contrary of being, that is as "not-being", and interpreted as lacking any form. Since logical concepts are forms, in the Western tradition no attempt was made to formally or conceptually construct, that is, inductively construct "nothing". But there are many examples of such inductions in Oriental philosophy, from which one should learn to see "nothing" in absolute darkness, to hear it in absolute quietness, to smell it in absolute pure air, to taste it in tasteless things and to grasp it where there is no resistance to the hand at all. Language expresses these experiences adequately in words like "there is nothing to see (hear, feel, taste, etc.)". Logic should follow this path in inductions of the multiple species of "nothing" in the different domains of sensible experience. – In logic false propositions are said to assert "nothing". And contradictory propositions, which are held to be logically false, can therefore also only assert "nothing". But what this actually means is concealed by the traditional verbiage of "absurdity", the "alogon" of Aristotle.

1.5. See Porphyry's introduction to Aristotle's Organon, as also: Porphyrius: On Aristotle's Categories, ed. and comm. by S. K. Strange, London 1992. –

G. F. Hegel, Phänomenologie des Geistes. System der Wissenschaft I (Phenomenology of Spirit. System of science I), Bamberg-Würzburg 1807, new 6th ed. by J. Hoffmeister, Hamburg 1952, p. 82: "Das Allgemeine ist also in der Tat das Wahre der sinnlichen Gewißheit" (The universal is in fact the truth of what is grasped with reliable certainty by the senses).

1.6.1. See Léon Baudry, Lexique philosophique de Guillaume d'Ockham, Paris 1958, art. Induction, p. 119: "Aliquando universale quod debet induci habet pro subiecto speciem specialissimam et ad habendam cognitionem de tali universali frequenter sufficit inducere per unam singularem" (Sometimes the universal concept which is to be induced has as an instance a most special concept, and to have knowledge of such a universal concept it suffices frequently to induce it from one singular instance). – F. Bacon, Novum Organum Scientiarum sive Iudicia vera de Interpretatione Naturae, Frankfurt 1564, Part II Aphorismi de Interpretatione Naturae 105: "In constituendo autem Axiomate, forma Inductionis alia, quam adhuc in usu fuit, excogitanda est; Eaque non ad Principia tantum (quae vocant) probanda et invenienda, sed etiam ad Axiomata minora, et media, denique omnia. Inductio enim quae procedit per enumerationem simplicem, res puerilis est, et precario concludit, et periculo exponitur ab instantia contradictoria, et plerumque secundum pauciora quam par est, et ex his tantummodo quae praesto sunt, pronunciat" (In order to establish an axiomatic (most general) concept, one has to conceive of another form of induction than was used hitherto. And it (applies) not only to the (categorical) principles (as they are called), but also to lesser and middle axioms (or species) and finally to everything. An induction which proceeds by simple counting (of instances) is a childish venture and concludes in an unsafe way; it is exposed to contradictory instances and (proceeds) mostly on fewer than needed, and arrives at results only on the basis of those at hand (Coll. 313)). Note that the usual translations of Bacon's text falsely assume that Baconian induction should establish true axiomatic propositions! – The maxim "individuum est ineffabile" can be traced back to Aristotle's dictum that the individual as a primary substance, described by a proper name, cannot serve as a predicate in a proposition ("Id quod non est praedicabile de multis", in: Peri hermeneias / De interpretatione 7, 17a38). However, Aristotle never formulated it expressly in those terms. Scholastic logicians tried to explain its meaning. One example is William of Ockham, who in his nominalistic vein presupposes that individuals may be known by "apprehension" alone in a "confused act of knowledge" (confusa cognitione); that is, like every other infinite continuum of things (cognitione confusa possunt infinita cognosci), and not by "abstractive cognition". See William of Ockham, Philosophical Writings ed. by Ph. Boehner, Edinburgh and London 1957, 2. ed. 1959, p. 45. This theory of the infinite complexity of the individual entered the later school books. E. g. Daniel Wyttenbach in his 'Praecepta Philosophiae logicae', Halle 1794, p. 118 says: "Sunt vero multae res, quae definiri nequeant. Primum Individua non Definitione, sed Descriptione declarantur. Neque enim eorum tot et tales notae possunt enumerari, quot et quales non in aliud etiam Individuum cadere queant" (There are many things that can't be defined. First of all, Individuals are explained not by definitions but by descriptions. For one cannot enumerate their intensions, which are so numerous and such that they can't be the property of any other individual). Johann Heinrich Zopf calls this property "Incommunicabilitas" (Incommunicability) in his 'Logica enucleata oder erleichterte Vernunft-Lehre, darinnen der Kern der alten und neuen Logic, wie auch der Hermeneutic, Methodologie und Disputier-Kunst begriffen', Halle 1731, p. 41: "Wenn man die Substantias eintzeln nach einander betrachtet, so heißen sie Individua, oder Supposita, deren vornehmste Eigenschaft ist Incommunicabilitas, weil kein Individuum dem andern seine Natur, die es selbst vor sich hat, mittheilen kann" (If one considers the substances individually one after another, than they are called Individua or Supposita whose primary property is Incommunicabilitas, because no individual can communicate its proper nature to an other individual). J. W. Goethe cites the maxim "Individuum est ineffablie" in a letter to Lavater of the year 1780, asserting that he would like to deduce a whole world out of this maxim. Wittgenstein's famous dictum "Was gezeigt werden kann, kann nicht gesagt werden" (Tractatus 4.1212) is certainly an echo of the maxim.

1.6.2. See J. St. Mill, A System of Logic Rationative and Inductive, being a Connected View of the Principles of Evidence and the Methods of Investigation (1843), ed. by J. M. Robson in the Collected Works of J. St. Mill, vol. VII. Toronto-London 1973, book III, chapter 3, § 2, p. 312–313 on black swans as an instance contrary to "fifty centuries" of European experiences. Nota bene: It was by no means necessary to categorize the newly discovered black birds as "swans"; biologists and Mill could just as well have inductively constructed a new species of "black non-swans" and given it a new positive denomination such as "swues".

1.7. On the various theories of numbers and number definitions see Chr. Thiel, art. "Zahlbegriff", in J. Mittelstraß (ed.), Enzyklopädie Philosophie und Wissenschaftstheorie, vol. 4, Stuttgart-Weimar 1996, p. 809–813; St. F. Barker, art. "Number", in: P. Edwards (ed.), The Encyclopedia of Philosophy, vol. V, New York 1967, p. 526–530; R. Knerr, Goldmann Lexikon Mathematik, art. "Zahlen", "Zahlengerade" and "Zahlensysteme", Gütersloh-München 1999, p 531–553. – This second proposal of B. Russell for a "definition" of numbers in his Introduction to Mathematical Philosophy, London 1919 (after the first proposal in the Principles of Mathematics, 1903, as "class of all classes similar to the (defining) class", which he later found to be paradoxical) now seems to be undisputed textbook wisdom. But obviously it is not a definition at all.

1.8. To relate induction to truth values, as J. St. Mill did, leads to "paradoxes of induction", as was shown by C. G. Hempel, N. Goodman und H. E. Kyburg. See L. J. Cohen, Inductive Logic 1948–1977, in: E. Agazzi (ed.): Modern Logic – A Survey. Historical, Philosophical, and Mathematical Aspects of Modern Logic and its Applications (Synthese Library 149), Dordrecht-Boston-London 1981, p. 354–356.

1.8.1. Mathematicians maintain that mathematical induction is paradigmatically "complete induction", because what holds for one number should also hold for all other numbers. But has any mathematician ever counted and does he know all numbers, and if not, how can he affirm anything about *all* of them? At best he knows an algorithmic procedure for constructing numbers, but certainly not all constructible numbers as such. See T. A. Skolem, Über die Nicht-Charakterisierbarkeit der Zahlenreihe mittels endlich oder abzählbar unendlich vieler Aussagen mit ausschließlich Zahlenvariablen, in: Fundamenta Mathematicae (Warschaw), 23, 1934, p. 150 – 161, repr. in: Selected Works in Logic, ed. by J. E. Fenstad, Oslo-Bergen-Tromsö 1970, p. 355–366. – The other side of that coin is that mathematicians and logicians believe logical induction to be paradigmatically "incomplete", because nobody could know all instances falling under an induced concept. But this can only be the case when the inductively constructed concept has no distinct intensions and is therefore not a genuine concept at all. Properly abstracted intensions show by themselves whether a newly detected instance has these intensions and falls under the induced concept or not. And this amounts to complete induction. See also 1.2.

1.8.2. The later Wittgenstein deserves great credit for spreading awareness of the fact that concepts can be unclear and indistinct, as his conception of the "family-resemblance of concepts" shows. But in the communities of students of the liberal arts and of fuzzy-logicians, what is in fact a deficiency of certain concepts has become the favorite view of what logical concepts as such actually are. Unfortunately, Wittgenstein confused homonymous words with genuine logical concepts. This certainly promotes creativity but is also pernicious for exact thinking. – See: R. Goeres, Die Entwicklung der Philosophie Ludwig Wittgensteins unter besonderer Berücksichtigung seiner Logikkonzeptionen, Würzburg 2000, chapter III/B3 on "The central conception of family-resemblance as the specific character of the concept", 234–300; also L. Geldsetzer, Wittgensteins Familienähnlichkeitsbegriffe, in: Internet HHU Duesseldorf 1999.

1.8.4. Quantification is now commonly considered as a mathematical and therefore exact determination. One thinks that the logical "all" and "some" mean exact numbers, "a" (or "one") means the number 1 (or some other mathematical "element"), and "none" means zero. But this is not the case, as will be shown below. The logical quantifications as introduced by Aristotle maintain their purely logical character as connectors between more general and more specific concepts. And they do so in mathematics itself, in addition to numerical quantification. – That variables are the mathematical counterparts of underdetermined concepts shows itself in the need to define them by equations.

1.8.5. W. V. O. Quine, Word and Object, Cambridge, Mass. 1960, chapter 2 on "Translation and Meaning" p. 26–79, espec. p. 29–33 on "gavagai". "Gavagai" serves obviously as an example for what Quine takes to be a concept. But if so, it can only be an unclear and indistinct concept. But the question is: are words concepts? – See also J. R. G. Williams, Gavagai again, in: Synthese 164, 2008, p. 235–259; L. Geldsetzer, Wörter, Ideen und Begriffe. Einige Überlegungen zur Lexikographie, in: Chr. Strosetzki (ed.), Literaturwissenschaft als Begriffsgeschichte (Archiv für Begriffsgeschichte, Sonderheft 8), Hamburg 2010, p. 69–96.

1.9.1. On Epicurus see Diogenes Laertius, Lives of Eminent Philosophers, ed. by R. D. Hicks, in The Loeb Classical Library, 1965, vol. 2, p. 623. Epicurus says: "One must not be so much in love with the explanation by a single way as wrongly to reject all the others". – Besides Leibniz' "Monadologia" see also J. M. Chladenius, "Einleitung zur richtigen Auslegung vernünftiger Reden und Schriften" (Introduction to the right exposition of reasonable speeches and writings), Leipzig 1742, repr. in: Instrumenta Philosophica Series Hermeneutica V, ed. by L. Geldsetzer, Düsseldorf 1969, p. 382.

1.10.1. Logicians commonly invoke the "Dictum de omni et nullo" which derives from Aristotle (Categories 3, 1b 10) and runs: "Quidquid de omnibus valet, valet etiam de quibusdam et singulis; quidquid de nullo valet, nec de quibusdam vel singulis valet" (What holds for all, holds also for some and for individuals; and what holds for none, does not hold for some or individuals). However this can only be said of generic intensions which are also intensions of their species and individuals, and which if absent from the genus are also absent from their subordinate species and individuals. This certainly holds for Aristotelian quantified syllogisms as also for numerical quantified expressions. As a maxim to deduce particular and individual propositions from a general proposition it would obviously result in false deductions. E. g.: "If all animals are living beings", it would be false to assert that (only) "some (or a definite number of) animals or (only) one individual animal is a living being". I daresay that many unaccountable errors in deduction stem from the misuse of this dictum.

1.10.2. See B. Ganter and R. Wille, Formale Begriffsanalyse. Mathematische Grundlagen, Berlin 1996. Engl. transl.: Formal Concept Analysis. Mathematical Foundations, Berlin 1999; B. Ganter and R. Wille, Begriffliche Wissensverarbeitung. I. Grundlagen und Aufgaben, Mannheim 1994; II. Methoden und Anwendungen, Berlin 2000.

1.10.5. See I. Kant, Critique of Pure Reason A 599/B 627.

1.10.6. See E. Cassirer, Substanzbegriff und Funktionsbegriff. Untersuchungen über die Grundlagen der Erkenntniskritik, Berlin 1910, 2. ed. 1923, repr. Darmstadt 1976, 5. ed. 1980. Engl. ed. Chicago-London 1923. Cassirer aimed at establishing "functional concepts" as the adequate logical form of mathematical and physical concepts and said that "the universal validity (universelle Gültigkeit) of a principle of series-construction (Reihenprinzip) is the characteristic of this concept" (1910, p. 26). But his explanation about how this works was far too vague to produce an impact on theories regarding the formation of logical concepts. See E. Horn, Der Begriff des Begriffs. Die Geschichte des Begriffs und seine metaphysische Deutung, München 1932, p. 56–58. – On quantitative or metrical concept formation see: W. Stegmüller, Probleme und Resultate der Wissenschaftstheorie und Analytischen Philosophie, Vol. 2: Theorie und Erfahrung (Problems and Results of the Philosophy of Science and of the Analytical Philosophy, vol. 2: Theory and Experience), Berlin–Heidelberg New York 1970, p. 44–109. – For an ultra-critical view on physical concepts see J. Marinsek, Rationale Physik oder wissenschaftliche Science Fiction? (Rational physics or scientific science-fiction?), Graz 1989.

1.11.1. See B. Russell, On Denoting, in: Mind 1905, repr. in B. Russell, Logic and Knowledge, Essays 1901–1950, ed. by R. C. Marsh, London 1965; G. Frege notably identified "functions" with "concepts" and "extensions of concepts" with the "truth values of functions": "Es erscheint zweckmäßig, Begriff geradezu eine Funktion zu nennen, deren Wert immer ein Wahrheitswert ist" (it seems appropriate to call 'concept' a function whose value is always a truth value). See G. Frege, Grundgesetze der Arithmetik, Begriffsschriftlich abgeleitet, vol. 1, Jena 1893, p. 7. Frege introduced this Cartesian and Leibnizian conception of true or false concepts into mathematical logic. And this signifies one of the most remarkable differences between mathematical logic and classical logic, where concepts as such can't be true or false.

1.12.3. On relational logic see I. M. Bochenski, Formale Logik, 3. ed. Freiburg-München 1970, p. 434–448; P. Geach, and G. H. von Wright: On an Extended Logic of Relations, Helsinki 1952; H. Höffding, Der Relationsbegriff. Eine erkenntnistheoretische Untersuchung, Leipzig 1922; R. P. Horstmann, Ontologie und Relationen. Hegel, Bradley, Russell und die Kontroverse über interne und externe Beziehungen, Königstein 1984; B. van Fraassen, Meaning Relations among Predicates, in: Nous 1, 1967, p. 161–89; B. van Fraassen, Meaning Relations and Modalities, in: Nous 3, 1969, p. 155–168. – For an example of pyramidal construction of the model concept of "father of ...", "grandfather of ..." see L. Geldsetzer, Logik, Aalen 1987, p. 106.

1.14. Conceptual contradiction is commonly an object of horror in epistemology and logic. J. F. Herbart, who propounded the thesis that the principal aim of the scientific enterprise is the elimination of unaccounted for conceptual contradictions, calls conceptual contradiction the "impossibility of a thought" (Unmöglichkeit eines Gedankens). See his Hauptpunkte der Metaphysik, Göttingen 1806 and 1808, p. 6. This is certainly the majority view to this day in all scientific communities and especially in mathematical logic and mathematics.

1.14.1. See L. Geldsetzer, Über das logische Prozedere in Hegels Phänomenologie des Geistes (On the logical procedure in Hegel's Phenomenology of Spirit), in: Jahrbuch für Hegelforschung 1, 1995, p. 43–80.

1.14.3. The received modern conception of the "possible" descends from Christian Wolff's definition: "Possibile est quod nullam contradictionem involvit" (The possible is that which does not contain any contradiction) (Philosophia prima sive Ontologia, Frankfurt 1729, § 85) or "was nichts Widersprechendes in sich enthält" (what does not include something contradictory) (Vernünftige Gedanken von den Kräften des menschlichen Verstandes, Halle 1712, I, § 12), and from his corresponding definition of the "impossible": "Impossibile dicitur quicquid contradictionem involvit" (The impossible is called that which contains something contradictory) (Philosophia prima sive Ontologia, § 79). Through Kant's talk of "conditions of possibility" this conception was transmitted into the foundations of current philosophy of science. I daresay that such an understanding of the possible and the impossible is, notwithstanding its noble ancestry, false and misleading.

1.14.4. That possibility can't be negated is the reason for the fact that so many asserted "impossibilities" have hitherto been demonstrated to be "possible" or even already realized. Hence the popular saying: "Never say 'impossible'" and Toyotas's advertising slogan "Nothing is impossible".

1.14.8. Modal logic has been a branch of logic since Aristotle and has developed into a highly specialized discipline. Its main conceptions are "necessary", "possible" and "factual" (or "real"), all of which are intertwined with the problem of "probability" or "verisimilitude". It makes sense to speak of the "necessity" of historical situations (which are assumed to be immutable), of "factual" when speaking about present objects, and of "possible" when conjecturing about future events (the "possibilia futura"), as Aristotle and the scholastics did. Leibniz introduced "possible worlds" into modern ontology by saying that God (out of his goodness) had created the real world as a choice of "the best of all possible worlds". And this idea remained the background for logical speculation about the characteristics of all this. See for the history: U. Nortmann, Modale Syllogismen, mögliche Welten, Essentialismus. Eine Analyse der aristotelischen Modallogik, Berlin 1996; K. Segerberg, Essay in Classical Modal Logic, 3 vols, Uppsala 1971; Carnap, R.: Meaning and Necessity. A Study in Semantics and Modal Logic, Chicago-Toronto-London 1947, 2. ed. 1956; German transl.: Bedeutung und Notwendigkeit. Eine Studie zur Semantik und modalen Logik, Wien-New York 1972. – On the current state of the art see A. Chagrov and M. Zakharyaschov, Modal Logic, Oxford 1997; P. M. Blackburn, M. de Rijke and Y. Venema, Modal Logic, Cambridge 2001. – Special contributions: J. Hintikka, Models for Modalities. Selected Essays, Dordrecht 1969, 2. ed. 1975; K. A. Bowen, Model Theory for Modal Logic. Kripke Models for Modal Predicate Calculi, Dordrecht 1978; L. Aquist, Modal Logic with Subjunctive Conditionals and Dispositional Properties, Uppsala 1971; D. Lewis, On the Plurality of Worlds, Oxford 1986; M. J. Loux, The Possible and the Actual, Ithaca 1979; B. Hale, Modal Fictionalism. A Simple Dilemma, in: Analysis 55, 1995, p. 63–67; T. Williamson, Bare Possibilia, in: Erkenntnis 48, 1998, p. 257–273. – It should be clear that, as a consequence of our view of the matter, the whole of modal logic is deviant and superfluous. "Necessary" may be used in a certain honorific way in describing logical demonstrations, "factual" as a characterization of application of logical forms to reality, and "possible" as a purely contradictory expression of imaginations and fantasies.

1.14.9. For a survey of the problem of dispositional predicates, which were introduced by R. Carnap, see W. Stegmüller, Probleme und Resultate der Wissenschaftstheorie und Analytischen Philosophie, Vol. 2: Theorie und Erfahrung (Problems and Results of the Philosophy of Science and of the Analytical Philosophy, vol. 2: Theory and Experience), Berlin-Heidelberg-New York 1970, p. 213–218; also W. Malzkorn, Defining Disposition Concepts. A Brief History of the Problem, in: Studies in History and Philosophy of Science 32A, 2001, p. 335–353; L. Geldsetzer, Logik, Aalen 1987, p. 94–98.

1.15. The view that numbers could be contradictory objects of thought seems to be a perennial taboo of arithmetic and a horror for mathematicians. Therefore the alleged proofs of their freedom from conceptual contradictions are famous, as for example G. Gentzen's "Die Widerspruchsfreiheit der reinen Zahlentheorie" (The Freedom from Contradiction (Consistency) of Pure Number Theory), in: Mathematische Annalen 112, 1936, p. 493–565, repr.

Darmstadt 1967; P. Lorenzen's "Die Widerspruchsfreiheit der klassischen Analysis" (The Freedom from Contradiction (Consistency) of Classical Analysis), in: Mathematische Zeitschrift 54, 1951, p. 1–24. – It seems that the belief that numbers present themselves in a natural order of "lesser" and "greater" derives from their age-old geometrical demonstration on a straight line where they are arranged in this way. This ordering inspired Descartes' analytical geometry and also Dedekind's definitions of real and irrational numbers. See J. W. R Dedekind, Stetigkeit und irrationale Zahlen, Braunschweig 1872, 7. ed. 1969, and: Was sind und was sollen die Zahlen? Braunschweig 1888, 10. ed. 1965, Engl. transl. "Continuity and Irrational Numbers" and "The Nature and Meaning of Numbers" in: R. Dedekind, Essays on the Theory of Numbers, Chicago 1901, repr. New York 1963. See also W. Sieg and D. Schlimm, Dedekind's Analysis of Numbers. Systems and Axioms, in: Synthese 146, 2005, p. 121–170. – Taken as objects of pure thought, numbers can only be called great or small in a metaphorical sense.

1.15.2. Deducing the concept of number by fusing quantifying logical connectors is an example of an effective deduction of a mathematical concept from logical presuppositions, as prescribed by Frege and Russell in their "logicistic program".

1.15.6. This can easily be seen in the numerical quantification of variables representing numbers. For example, if the variable a represents the number 5, one may logically say that "3a" means "three fives", and that ends the matter. But in arithmetic one is compelled to learn that the quantification of a number means another number, for example, "3a (where a=5)=3 · 5=15". The extensional quantification by numbers is fused (through multiplication) with an intensionally characterized number, which results in yet another number.

1.15.7. Leopold Kronecker in "Ueber den Zahlbegriff", in: Journal für reine und angewandte Mathematik 10, 1887, p. 261–274, also in: L. Kronecker, Werke, vol. 3, ed. by K. Hensel, Leipzig 1899, repr. New York 1968, notably introduced the "natural numbers" (in contrast to G. Cantor, K. Weierstraß and R. Dedekind) with a theological argument: "Die (positiven) ganzen Zahlen hat der liebe Gott gemacht, alles andere ist Menschenwerk" (the natural numbers were created by the good Lord, but the rest are man-made). Which was an easy way not to explain them. See also: Ch. Thiel, art "Zahlbegriff" in: J. Mittelstraß (ed.), Enzyklopädie Philosophie und Wissenschaftstheorie, vol. 4, Stuttgart-Weimar 1996, p. 809–813; H. Hodes, Where Do the Natural Numbers Come from?, in: Synthese 84, 1990, p. 347–407.

1.15.8. B. Russell's discovery of the paradox of "the class of all classes" in his "Principles of Mathematics" of 1903 (see B. Russell, Les paradoxes de la logique, in: Revue de métaphysique et de morale 14, 1906, p. 627–650, Engl. transl. as: On 'Insolubilia' and their Solution by Symbolic Logic, in: B. Russell, Essays in Analysis, ed.

by D. Lackey, London 1973) was received with shock in the mathematical profession and triggered a host of further detections of paradoxes and antinomies as well as much work to "dissolve", "defeat" or simply "forbid" them (as Russell's "theory of types" proposed). In fact the class paradox is based on the reflexive use of the concept of class or set. Apparently, mathematicians have not yet learned to live with and use paradoxes as isolated phenomena in their theories. – See also A. De Morgan, A Budget of Paradoxes, 2 vols, ed. by S. de Morgan, London 1872, 2. ed. 1915, repr. New York 1969; P. Geyer and R. Hagenbüchle (eds.), Das Paradox. Eine Herausforderung des abendländischen Denkens, Tübigen 1992; R. M. Sainsbury, Paradoxes, Cambridge 1988, 2. ed. 1995.

1.15.9. On Georg Cantor's achievements see O. Becker, Grundlagen der Mathematik in geschichtlicher Entwicklung (Foundations of Mathematics in its Historical Development), 2. ed. Freiburg-München 1964, p. 277–314. – The similarity which Cantor proclaimed between the highest arithmetical totality or "Mächtigkeit" and God's omnipotence, in the fashion of Nicholas Cusanus, did not amuse all mathematicians.

1.16. Compare this pyramid of the number-concepts with the mere classifications of numbers in: W. and M. Kneale, The Development of Logic, 3. ed. Oxford 1964, p. 394, or in R. Knerr, Goldmann Lexikon Mathematik, Gütersloh-München 1999, p. 337.

1.16.1. In the history of mathematics each construction of a new kind of number has been and still is considered a decisive progress in its development. Even today, some of these still lack a formal definition by an adequate algorithm, as typically shows itself in the prime numbers. – On the non-standard numbers see A. Robinson, Non-Standard Analysis, Amsterdam 1966, repr. 1974; A. Robinson, Selected Papers, ed. by H. J. Keisler a. o., Amsterdam 1979.

1.16.3. That Euclid distinguished between, on the one hand, division into equal halves and other partitions on the other can be seen from his express mention of each. See Euklid, Die Elemente, Buch I-XIII, ed. and transl. into German by Clemens Thaer, Darmstadt 1962, book 7, sect. 6: "an even number can be divided into halves", and sect. 7: "an odd number is not divisible into halves". On Euklid's mathematical endeavours see Moritz Cantor, Vorlesungen über Geschichte der Mathematik, vol. 1, 3. ed. Leipzig 1907, p. 258–294.

1.17. Prime numbers in arithmetic are like comets in astronomy: they do not seem to obey laws. This renders them useful for military, diplomatical and industrial ciphering. However, their supposed lawlessness derives first from the fact that mathematicians take the number 2 to belong to them while excluding the "one", and secondly from the age-old dogmatic habit of mathematicians, following Euclid, of believing them to be lawless. Our logical viewpoint may aid in deciphering the primes' behaviour. Remember: The neo-platonists Nikomachos and Jamblichos both eliminated the number 2 from the primes. See Cl. Thaer in his Euklid-edition, annotations to book 7, Def. 11, p. 439; and Moritz Cantor, Vorlesungen über Geschichte der Mathematik, 3. Aufl. Leipzig 1907, p. 461. And later there were mathematicians, such as Henry Lebesgue, who took the 1 to be prime. The reason why the number 1 has generally not been accepted as prime is obviously due to the fact that Euclid notably did not consider the "one" (as the "unity") to be a number at all. – On prime numbers generally see Wl. Narkiewicz, The development of prime number theory. From Euclid to Hardy and Littlewood, Berlin 2000, as well as the articles in Wikipedia.

1.17.2. In previous publications as in Logik, Aalen 1987, p. 154, as well as in "Grundriß der pyramidalen Logik", Internet HHU Duesseldorf 2000) I used the equivalent formula "$(2x+1) \cdot (2y+1)$ (for $x = 1,2,3,4\ldots$; $y = 1,2,3,4\ldots$)".

1.18. See J. F. Fries, Platons Zahl, De Republica L. 8, p. 546 Steph. Eine Vermutung, 1823. Also in: J. F. Fries, Sämtliche Schriften, ed. by G. König and L. Geldsetzer, vol. 20, Aalen 1969, p. 355–414. It seems that philologists have not yet taken notice of this result.

2. Aristotle distinguished "categorical terms" (such as subject and predicate) from "syncategorical terms" in propositions. But he never explained what the latter are. Only the scholastics later provided such an explanation. See, e.g. William of Ockham's definition: "syncategorema proprie loquendo nihil significat, sed magis additum alteri facit ipsum significare aliquid, sive facit ipsum pro aliquo vel aliquibus aliquo modo determinato supponere, vel aliud officium circa categorema exercet" (a syncategorematic term does not signify anything, properly speaking; but when added to another term, it makes it signify something or makes it stand for some thing or things in a definite manner, or has some other function with regard to a categorematic term). See William of Ockham, Philosophical Writings, ed. by Ph. Boehner, Edinburgh and London 1957, 2. ed. 1959, p. 51. L. Wittgenstein continues this tradition: "Mein Grundgedanke ist, daß die 'logischen Konstanten' nicht vertreten" (My fundamental idea is that the logical constants, i. e. connectors, do not represent (as names represent things)). See Tractatus logico-philosophicus / Logisch-philosophische Abhandlung (1921) 4.0312, Frankfurt a. M. 1963, p. 37.

2.1. The reason for this neglect of expression-forming connectors obviously lies in the way in which truth value tables of propositional logic have defined all logical connectors as "truth value bearers" of connected "elementary sentences", following therein the examples of the Stoics. See I. M. Bochenski, Formale Logik, 3. ed. Freiburg 1970, p. 133–140, as also H. Lenk, Kritik der logischen Konstanten. Philosophische Begründungen der Urteilsformen vom Idealismus bis zur Gegenwart, Berlin 1968.

2.2. The result of pyramidal formalization is what Wittgenstein postulated as an ideal but did not himself achieve, namely that one can read off from the formalism itself whether the propositions are true or false. Tractatus logico-philosophicus 6.113, p. 94: "Es ist das besondere Merkmal der logischen Sätze, daß man am Symbol allein erkennen kann, daß sie wahr sind, und diese Tatsache schließt die ganze Philosophie der Logik in sich" (It is the special characteristic of logical propositions that one can read off from the symbolism alone that they are true, and this fact includes the whole of philosophy of logic).

2.2.1. See L. Wittgenstein, Tractatus logico-philosophicus (1921) 5.101, and Emil Leon Post, Introduction to a general theory of elementary propositions, 1921 (Diss. of 1920).

2.2.2. Remember what was said about "meta-" and "meta language" in 0.6.5. and 0.6.6. On deficiencies deriving from the missuse of "meta-sense" see below 4.6.6. to 4.6.8.

2.2.3. See, e.g., the proposed denominations of the 16 connectors in I. M. Bochenski and A. Menne, Grundriß der Logistik, 4. ed. Paderborn 1973, p. 35.

2.2.5. Concerning the inadmissible use of implication for representations of identity ("if p, then p") see 7.1.7. – One may doubt whether Wittgenstein meant what he formalized. But compare what he asserted himself in 6.54 of his Tractatus, namely that his sentences are "unsinnig" (the German word oscillates between "absurd" and "senseless").

2.3.1. See L. Wittgenstein, Tractatus logico-philosophicus 5.101, p. 60: "(WWWW) (p.q) Tautologie (Wenn p, so p; und wenn q, so q) (p<p . q<q)". – The false "connection" of something with itself was called by the Stoics an inadmissible "repetition" of a name in logical reasoning. Later on it entered into the definition of "substance" (as a kind of "relation") in Kant's table of categories. It remains the foundation of all "reflexive" thinking.

2.4.1. This ambiguity of the Aristotelian "attribution" (German: "Zukommen") seems to be the reason that "inclusion" and "implication" as well as "belonging to" are frequently understood in this double sense.

2.4.4. See Sextus Empiricus, Pyrrhoneische Grundzüge, übers. und hgg. von Eugen Pappenheim, Leipzig 1877, III, 3, S. 170: "Das Ursächliche muß entweder mit der Wirkung zusammen bestehen, oder vor dieser bestehen, oder nach ihr werden. Zu sagen nun, daß das Ursächliche ins Bestehen geführt werde nach dem Werden seiner Wirkung, – daß das nur nicht sogar lächerlich ist! Aber auch vor dieser kann es nicht bestehen; denn in Bezug auf sie wird es gedacht, wie man sagt ... Aber (es kann) auch nicht (mit der Wirkung) zusammen bestehen" (The cause must either exist together with the effect, or before it, or come forth after it. Now, to say that the cause came into existence after its effect, is ridiculous! Neither can it exist before it (the effect), because it is imagined (i. e remembered) in relation to the effect, as one says. Nor can it exist together (with its effect)). – Nota bene: the "imagined" (or remembered) concept of the cause is just as much a concept as is the concept to which the immediate sensory experience of the effect gives rise.

2.5.2. See further J. L. Garfield: The Fundamental Wisdom of the Middle Way. Nagarjuna's Mulamadhyamaka karika, Translation (from the Tibetan) and Commentary, New York-Oxford 1995; Nagarjuna, Die Lehre von der Mitte (Mula-madhyamaka-karika) Zhong Lun, Chinesisch-Deutsch, translation from the Chinese version of Kumarajiva with commentary by L. Geldsetzer, Hamburg 2010, p. 146–148 and p. 152.

2.6.1. Meinong's paradox was exemplified by the proposition "there is no golden mountain", which might have been simply false when one considers that there is a mountain of gold bricks at Fort Knox, USA. And this also holds in the case of B. Russell's "present king of France", at least in the view of recent French monarchists. Taking these concepts as dispositional concepts (like Pegasus), it is obvious that one cannot experience real mountains of gold nor kings of France in the present, but has to combine the separate experiences of gold and mountains and of historical French kings and the recent republican era in fantasy. Then one observes that those concepts have the extensions of their components. See Alexius v. Meinong, Über Gegenstandstheorie. English translation by I. Levi: "The Theory of Objects", in: R. M. Chisholm (ed.), Realism and the Background of Phenomenology, New York 1960. See also K. J. Perszyk (ed.): Nonexistent Objects. Meinong and Contemporary Philosophy, Dordrecht 1993.

2.7. This pyramid of all connectors has been constructed logically on the basis of what has been previously said regarding concepts and their interrelations. Compare it with L. Wittgenstein's rather rhapsodic "truth value tables" for 16 connectors (Tractatus logico-philosophicus 5.101, p. 60). As is well known, "truth value tables" have entered almost all textbooks of logic, and, with respect to that half of the connectors which do not resemble classical logical connectors and whose functions are therefore most conspicuous, students are left free to consider what the fantastic names which have been given to them might mean. See, e. g., I. M. Bochenski and A. Menne, Grundriß der Logistik, 4. ed. Paderborn 1973, p. 35.

2.8. The discovery and establishment of such equivalences between connectors, which are often called "laws", occupies an obligatory chapter in textbooks on mathematical logic. See f. i. I. M. Bochenski and A. Menne, Grundriß der Logistik, 4. ed. Paderborn 1973, p. 38–66. These equivalences are mainly used in the transformations of equations within proofs.

2.9. Quantification has been and still is held to belong to the realm of propositional assertions. This is signaled by the traditional use of subalternation (see 1.10.1.) as well as by its absence from the definitions contained in truth value tables (see 2.2.4.). But to "affirm" quantified propositions presupposes completely or incompletely defined concepts!

2.9.1. G. Boole in "An Investigation of the laws of thought, on which are founded the mathematical theories of logic and probability", London 1854, replaced the logical "none" by the mathematical "zero" and created thereby the prototype of empty concepts in mathematical logic. See I. M. Bochenski, Formale Logik, 3. ed. Freiburg 1970, p. 353 f., as also Y. Balshov, Zero-Value Physical Quantities, in: Synthese 119, 1999, p. 253–286.

2.10.1. Ch. S. Peirce apparently contributed to the "merger" of equivalence and copulative assertion by inventing a special logical sign "€", which combined "C" (for subordination, which stands for the copula, e.g., "Gold is Metal") and "=" which expresses the equation (Salt=Sodium chloride). He says: "Die Kopula ‚ist' wird bald die eine, bald die andere der beiden Beziehungen ausdrücken, die wir mittels der Zeichen C und = dargestellt haben. ... Ausführlichst wird dieses Zeichen als ‚untergeordnet oder gleich' zu lesen sein". (The copula "is" will express now one, now the other of the two connections which we have represented by means of the signs "C" and "=" ... This sign is to be most fully read as ‚is subordinate or equal to'.) See Ch. S. Peirce, Vorlesungen I, citation from I. M. Bochenski, Formale Logik, p. 357.

2.10.2. Leibniz' view has entered mathematical logic via the definition of equivalence in Wittgenstein's tables, where equivalences are said to have truth values, namely: If two propositions linked as equivalent are both true or both false, than the equivalence is true; if two propositions have different truth values, than the equivalence is false. See L. Wittgenstein, Tractatus logico-philosophicus 5.101, p. 60. Textbooks tell us that this equivalence may be read as strict implication: "If p, then and only then q". Consider one example: "If equivalences have truth values, then and only then may all true propositions substitute for one another 'salva veritate', and all false propositions also substitute for one another 'salva veritate' ". – Nota bene: one may speak of truths and falsehoods and connect them at will. But then they are not the "meta-sense" of the combined propositions.

2.10.5. In mathematics negated equalities are understood as inequalities and widely used in the formula "x≠y". But dialectical mathematical thinking is at work here, too: Inequalities can also be understood as defining equalities. "x≠y" means also "x=not y", that is: The numerical values represented by x are all those numerical values not represented by y.

2.11. It seems that mathematicians have never considered the ways in which the connectors used in calculations to construct sums, differences, quotients, products, integrals etc. actually operate to combine numbers into number expressions. David Hilbert proposed the easy way to avoid this task by making those mathematical connectors into axioms, which require no explanation. See D. Hilbert, Über den Zahlbegriff' (1900) in: O. Becker, Grundlagen der Mathematik in geschichtlicher Entwicklung, 2. ed. Freiburg-München 1964, p.353 f. – G. Boole introduced the commonly accepted – but highly misleading – expression, that the disjunction is a "logical sum", and the adjunction a "logical product".

2.12. The mathematical belief that equivalences are genuine (and true) propositions has survived all mathematical foundational crises and clearly marks the difference between mathematical and logical thinking. See, e. g. H. Weyl, Über die Grundlagenkrise der Mathematik (On the foundational crisis in mathematics), in: O. Becker, op. cit. p. 350. Weyl maintains: "(Beispiel für) ein wirkliches Urteil (ist) $17+1=1+17$" ((An instance) of a real judgment (is) $17 + 1 = 1 + 17$), whereas "'es gibt eine gerade Zahl' – ist überhaupt kein eigentliches Urteil im eigentlichen Sinne, das einen Sachverhalt behauptet" ('there is an even number' – is not a proper judgment which asserts a state of affairs)!

2.13. "$\surd 4=+2$ and -2" defines two meanings of the expression "$\surd 4$". Taken as a propositional assertion it would result in the manifest contradiction "$+2=-2$" or in common speech: "plus 2 is minus 2".

2.14. This way of defining kinds of numbers proceeds by violating and contradicting the previously existing valid definitional modes. Almost every proposal for definitions of new number-kinds was felt as "contradictory" and rejected by one mathematician or another. E. g. M. Stifel in his "Arithmetica Integra", Nürnberg 1544, defined the negative numbers as "numeri ficti infra nihil" (fictitious numbers below, that is: smaller then nothing)! And François Viète (1540–1603) never acknowledged them as numbers at all. Descartes called "even roots" of negative numbers "imaginary" in contrast to "real numbers" because he could "not imagine" what their meaning should be. Later mathematicians "imagined" them in Gauss' geometrical fashion and retained the name.

2.14.1. Addition and subtraction in the positive range are synonymous with logical adjunction and negated adjunction. Multiplication is synonymous with fusion of concepts. That is why all three operators apply to concepts generally, and are not restricted to numbers.

2.15.3. Charles Bouillé (Bovillus, Bouvelles, 1470 – ca. 1553), in the manner of the Lullian art, tried to introduce mathematical powers into logic. He defined the first man Adam as "homo", his wife Eve as "homo -homo" und their first son Abel as "homo-homo-homo". See Carolus Bovillus, Liber de Sapiente, in the Reprint of the edition of his works (Paris 1510), Stuttgart-Bad Cannstatt 1970 (appeared only 1973) p. 132. The method has to some extent been followed in the self-reflexive formation of concepts, for example, "self-consciousness" (= consciousness-consciousness). – J. H. Lambert, one of the founders of the algebra of logic, formalized a general concept a as "$a=a\gamma+a\delta$", where a means the concept, γ its "genus proximum", and δ its "specific difference". The next higher concept which comprises a, was then formalized as "a ($\gamma+\delta)^{2}$". See C. Mangione, Logica e fondamenti della matematica, in: L. Geymonat (ed.), Storia del pensiero filosofico e scientifico, 3rd ed. Milano 1979, p. 142. – But if so, "fruit" should be defined as "apple^{2} ".

2.15.4. The confusion of these two kinds of quotients has had misleading effects in the probability calculus. The commutative proportion of "knowledge / ignorance" or "ignorance / knowledge" which characterizes conjectures is calculated by division as a part of the meaning of their expression and thus quantified by a number. But because of the commutativity of the proportion there result two "solutions" signifying parts of knowledge or parts of ignorance.

2.15.5. With the calculation of differential equations the student of mathematics enters the higher regions of dialectical "infinitesimal" thinking. It is, as honest teachers confess, "immer noch nicht völlig erforscht" (still not yet completely investigated), see R. Knerr, Goldmann Lexikon Mathematik, Gütersloh-München 1999, art. "Infinitesimalrechnung", p. 163. It probably never will be as long as no mathematician is inclined to concede that Leibniz' "differential quotient" as well as Newton's "motion at the beginning or the end of motion" are utterly contradictory concepts veiled in geometrical talk of "coincidence" of infinitesimal extensions of lines and points or in arithmetical verbiage of limitless approximation of the quotient 0/0. – George Berkeley did not so much "criticise" as "analyse" Newton's "infinitesimal fluxions" in his "The Analyst", London and Dublin 1734, as contradictory notions, saying: "they are neither finite magnitudes nor infinitely small ones, nor nothing. Should we not call them specters of departed magnitudes?" And he suggested – in the manner of Nicolas Cusanus – that "objects, principles and modes of inference of modern analysis" are not clearer nor more evident than "religious secrets" and dogmata. See O. Becker, a. a. o. p. 156–158. The same O. Becker asserts – as most mathematicians now do – "that we cannot any longer acknowledge those (arguments) as really valid" (p. 156).

2.16.1. Abstraction from geometrical (visible) content leads – via Cartesian "analytical geometry" – to purely arithmetical "analysis". The latter investigates the conditions and restrictions under which the relations between expressions which involve variables result in equations. Inspired by the model of the balance, Descartes wrote in his "Géometrie"of 1637: « (Les équations sont) des sommes composées de plusieur termes partie connus et partie inconnus dont les uns sont égaux aux autres, ou plutôt qui, considérés tous ensembles, sont égaux à rien: car ce sera souvent le meilleur de les considérer en celle sorte » / "(the équations) are sums composed by concepts partly known and partly unknown, either being equal to the other, so that taken together, they are equal to nothing : which is often the best mode of considering them". Since there is no limit to formulating ever more complex expressions and to setting them in equational relations, reduction to the unities has methodical limits. These limits have the effect that such "hyperkomplex" expressions are definitions of "transcendent" number expressions.

2.18. Using "possible" and "probable" as connectors has led (since Aristotle) to modal and and many-valued logics. Both are in fact mistaken attempts to transform conjectures into assertions. And this results in clothing wishes, hopes, dreams and speculations in the formalized concepts of "possible worlds". For a succinct view of the matter see I. M. Bochenski and A. Menne, Grundriß der Logistik, 4. ed. Paderborn 1973, p. 112–121.

2.18.1. This adoption of the indicative instead of the subjunctive mode is the basis of the traditional belief (since the Megarians) that the inference of falsehoods from falsehoods is true, as the example of Philo shows: "If the earth flies, then the earth has wings" (as a true "Philonic implication"). But no one would ever assert – unless lying or joking – that the earth flies or has wings, although one can always conjecture: "in the case in which the earth flew, it would have wings". The strict distinction between these modes of thinking has remained a legacy of the Stoics to modern jurisprudence.

3. D'Alembert was apparently the first and last to observe that mathematical equations are logical equivalences. In his introduction to the "Encyclopédie" of 1759 he rightly says: « Qu'est-ce que la plupart de ces axiomes dont la géométrie est si orgueilleuse, si ce n'est l'expression d'une même idée simple par deux signes ou mots différents? Celui qui dit que deux et deux font quatre, a-t-il une connaissance de plus que celui qui se contenterait de dire que deux et deux font deux et deux? » / "What are most of these axioms geometry is so proud of, if not the expression of one and the same simple idea by two different signs or words? If someone says that two and two make four, does he know more than someone who would say that two and two make two and two?", see: D'Alembert, Discours préliminaire de l'Encyclopédie, Paris 1965, p. 39. – Definitions are frequently confused with but also distinguished from propositions, depending on whether the meanings of the copula and the equals sign are confused or distinguished, as well as from the understanding of "induction". This shows up in relevant dictionaries, f. i. G. Gabriel, art. "Definition" in: J. Mittelstrass (ed.), Enzyklopädie Philosophie und Wissenschaftstheorie, vol. 4, Stuttgart 1996, p. 439–442. Generally speaking, mathematicians and mathematical logicians understand definitions as propositions with truth values formalized by equations (which however in fact are not at all truth-conducive!); classical logicians and philologists mostly distingish definitions from propositions but formalize the definitions with the copula (thereby assigning them the logical form of genuine copulative propositions). – For various views of the problem see R. Borsodi, The Definition of Definition. A New Linguistic Approach to the Integration of Knowledge, Boston, Mass. 1967; J. Winnie, The Implicit Definition of Theoretical Terms,

in: British Journal for the Philosophy of Science 18, 1967, p. 223–229; D. Lewis, How to Define Theoretical Terms ? In: Journal of Philosophy 67, 1970, p. 427–446; C. Peacocke, Truth Definitions and Actual Languages, in: Truth and Meaning, ed. by G. Evans and J. McDowell, Oxford 1976, p. 162–188; G. Gabriel, Implizite Definitionen. Eine Verwechselungsgeschichte, in: Annals of Science 35, 1978, p. 419–423; R. Kleinknecht, Grundlagen der modernen Definitionstheorie, Königstein 1979; G. Weaver, A Note on Definability in Equational Logic, in: History and Philosophy of Logic 15, 1994, p. 189–199; J. H. Fetzer, D. Shatz and G. Schlesinger (eds): Definitions and Definability: Philosophical Perspectives, Dordrecht 1991; S. Feferman, Definedness, in: Erkenntnis 43, 1995, p. 295–320; R. M. Francescotti, How to Define Intrinsic Properties, in: Nous 33, 1999, p. 590–609. – Frege was seriously mistaken in maintaining that equality refers to either truth or falsity and only to these "truth values". Notably, he called every equation "the name of a truth value". See G. Frege, Grundgesetze der Arithmetik, begriffsschriftlich abgeleitet, Vol. 1, Jena 1893, p. 6. He thereby transmitted the time-honored opinion of mathematicians that equations are concerned with truth or falsity. K. Ajdukiewicz in his "Abriss der Logik", Berlin-Ost 1958, p. 37 maintains: "Die wichtigste Bedingung, die die Definition zu erfüllen hat, ist ihre Wahrheit" / "The most important condition which the definition has to fulfill is its truth". He identifies the expression "das heißt soviel wie" (that means as much as) with the copula "ist" (is) and calls this hybrid idea the "Definitions-Bindeglied" (connector for definitions) (ibid. p. 35).

3.1.2. The definition A = B is the normal case in two-language dictionaries where different words express one and the same meaning. The definition A = non-B is rarely taken into account by logicians, because it looks somewhat perplexing. But it is also customarily used for definitions (f. i. by negated terms) in single language dictionaries.

3.1.3. Johann Gottlieb Fichte introduced this misuse into philosophy and draw misleading conclusions from it. He says "Den Satz A ist A (soviel als A = A, denn das ist die Bedeutung der logischen Kopula) gibt jeder zu; und zwar ohne sich im geringsten darüber zu bedenken: man anerkennt ihn für völlig gewiß und ausgemacht" / "The sentence A is A – which means A = A because this is the meaning of the logical copula – is acknowledged by everybody, and indeed without the least reflection: it is acknowledged as totally certain and confirmed", see J. G. Fichte, Grundlage der gesamten Wissenschaftslehre (1794), ed. by W. G. Jacobs, Hamburg 1970, p. 12. – It is revealing that mathematics and mathematical logic introduced special signs to express definition in order to distinguish it from equivalence. But since both have the same meaning this is – apart from the confusion of the copula with equivalence – another source of dialectical thinking (which can for the sake of clear thinking be cut off with Ockham's razor).

3.2. Aristotle produced definitions of concepts (Greek: horismos) by induction (epagogé) from singular instances. He presupposed (what was called in 1.9.1.) an inductive frame concept which transfered its "generic" intensions additionally to the specific intensions of the instance. His classical definition of the definition runs: "ho horismos ek genous kai diaphoron esti" / "The definition stems from the genos and the differences", see Aristotle, Topica I, 8, 103 b 15 f. – The Aristotelian type of definition is still in use in all sciences, also in the non-technical parts of mathematical and physical texts. Speakers and writers of cultured languages use it almost automatically "to make their ideas clear".

3.2.1. The customary use of the Aristotelian definition has led to the widespread belief that the next higher concept suffices for the definition of a concept. This may serve well enough in some contexts. But it is inadequate in principle, as pyramidal formalization shows. It diverts attention from the need to indicate the intensions deriving from the highest concepts or categories to make a definition complete.

3.3.1. See M. Fernández García, Lexicon Scholasticum philosophico-theologicum in quo termini, definitiones, distinctiones et effata a Joanne Duns Scoto exponuntur, declarantur (1910), repr. Hildesheim-New York 1974, p. 246 f., art. "Ens", "Entis divisio", and p. 700 f., art. "Univocum", "Univocum – aequivocum –analogum". – If "being" (Greek: on, Latin: ens) is the highest concept then its meaning enters all lower concepts within its scope as a "generic" intension.

3.4. There are good reasons to think that Kant's famous "conditions of possibility" ("Bedingungen der Möglichkeit") of concepts were constructed in the Thomist fashion.

3.4.1. See Aristotle, Metaphysics Book 4, 2, 1003a: "The term 'being' is used in various senses, but with reference to one central idea and one definite characteristic, and not as merely a common epithet", in: Aristotle, The Metaphysics, ed. by H. Tredennick, Cambridge, Mass. 1956, p. 147.

3.5.1. The Stoics seem to have understood that particular sentences are not assertions with truth values, judging from the fact that they never used particularisation in their logical arguments. It seems, however, that Aristotle's frequent use of particular (and individual) "propositions" in his syllogisms caused them to be confused with genuinely assertive propositions, a confusion which persists to this day.

3.5.2. The algorithm takes its name from the Persian mathematician Al-Chwarismi (ninth century A. D.), author of a famous mathematical textbook. See K. Vogel (ed.): Mohammed Ibn Musa Alchwarizmi's Algorismus. Das früheste Lehrbuch zum Rechnen mit indischen Ziffern, mit Transkription und Kommentar, Aalen 1963. Algorithms presuppose the use of variables standing for general number concepts. Deducing individual instances and particular samples of numbers or signs from algorithmes has been routinized by computers whose programs themselves have the structure of algorithms. However, the construction of an algorithm which defines given individual or particular mathematical or other signs is a matter of mathematical research. This is demonstrated by the fact that no algorithm has been found to date for what are concidered the known prime numbers. See: Wl. Narkiewicz, The development of prime number theory. From Euclid to Hardy and Littlewood, Berlin 2000. Mathematical logic discusses the relevant problems in terms of definability vs. undefinabiltity, completeness vs. incompleteness, decidability vs. undecidability, satisfiability vs. unsatisfiability. These relate to the intensional and extensional determinations of the algorithms and of their individual and particular deductions in the definitional equations. But all this has nothing to do with the (falsely) assumed truth or falsity of the involved definienda.

3.7.2. The introduction of letters into mathematics, called "logistica speciosa" (quae per species seu rerum formas exhibitur), is due to François Viète (Franciscus Viëta, 1540–1603). See F. Viëta, In Artem Analyticam Isagoge seu Algebra Nova, Tours 1591, French ed.: Introduction en l'art analytique, ou nouvelle algèbre de Francois Viète, Paris 1630; Engl. ed. in: J. Klein: Greek Mathematical Thought and the Origin of Algebra, Cambridge, Mass. 1968; German ed. by K. Reich and H. Gericke: F. Viète, Einführung in die neue Algebra, München 1973.

3.7.4. L. Wittgenstein, Tractatus logico-philosophicus, 6. 1., p. 93: "Die Sätze der Logik sind Tautologien" / "The propositions of logic are tautologies".

4. There are good reasons to look back to Johann Heinrich Lambert's "Neues Organon oder Gedanken über die Erforschung und Bezeichnung des Wahren und dessen Unterscheidung von Irrthum und Schein" (New Organon or thoughts concerning the inquiry into and the naming of the true, and how it is distinguished from error and appearance), 2 vols, Leipzig 1764, repr. Hildesheim 1965. – Much of the veneration and trust in the public's attitude toward sciences obviously derives from the fact that scientists like to claim that scientific knowledge is always and solely true.

4.1. In classical logic the truth value-bearing element of logic was called judgment or sentence (German: "Urteil" oder "Behauptungssatz"). This is now the field of predicate logic, so-called because propositions are understood as composed of a subject and a predicate concept. "Proposition" was introduced into mathematical logic to distinguish the meaning or the "thought" (Frege's "Gedanke") of a sentence or judgment from its written or spoken form. About Kant's (false) thesis and argument that mathematical equations are "propositions", and especially "synthetic a priori" (rather than logical equivalences), see his Prolegomena § 2, as also his Critique of pure Reason B 16. The specifically so-called propositional logic derives from B. Russell's "Principles" and L. Wittgenstein's definitions of truth values of combined "elementary sentences" as forming complex "propositions" which are similar to syllogistic arguments. Propositional logic is now considered almost synonymous with "modern" and "mathematical logic", at least by mathematicians and many philosophers of science. However, it does not merit this honor. Freed from purely mathematical presuppositions and foundational errors (which are criticized in this book) it can be reduced to the classical logic of judgments. Therefore we use the now customary term "proposition" (German: "Aussage") to denominate truth value-bearing logical elements in general. – See regarding the styling of the "discipline" H. A. Schmidt, Mathematische Gesetze der Logik (Vorlesungen über Aussagenlogik), Berlin-Göttingen-Heidelberg 1960; W. Rautenberg, Klassische und nichtklassische Aussagenlogik, Braunschweig-Wiesbaden 1979; P. Gochet, Outline of a Nominalist Theory of Propositions. An Essay in the Theory of Meaning and in the Philosophy of Logic, Dordrecht-Boston-London 1980; R. L. Epstein et al., The Semantic Foundations of Logic, Vol. 1: Propositional Logics, Dordrecht 1990, 2. ed. 1995; H. Kleine Büning and Th. Lettmann, Propositional Logic. Deduction and Algorithms, Cambridge 1999.

4.2. Particular propositions have been included since Aristoteles and throughout the scholastic era among the "categorical propositions" and symbolized in the syllogistic denominations of the syllogisms with special letters: a for general affirmative, i for particular affirmative, e for general negative and o for particular negative propositions. This presupposes that they have truth values. The categorical propositions also included the "individual propositions" which were mainly used in Aristotles' examples of syllogisms. Scholastics made much ado about their interrelations and worked out the famous "logical quadrat" or "square of oppositions" for this purpose. See still K. Ajdukiewicz, Abriss der Logik, Berlin-Ost 1958, p. 112–122.

4.4.1. L. Wittgenstein, Tractatus logico-philosophicus, 6.113, p. 94: "Es ist das besondere Merkmal der logischen Sätze, daß man am Symbol allein erkennen kann, daß sie wahr sind, und diese Tatsache schließt die ganze Philosophie der Logik in sich" / "It is the special characteristic of logical propositions that one can read off from the symbolism alone that they are true, and this fact includes within itself the whole philosophy of logic".

4.6.4. See the classical works on paradoxes such as B. Bolzano, Paradoxien des Unendlichen, ed. by F. Prihonsky, Leipzig 1851, repr. Darmstadt 1964; A. De Morgan, A Budget of Paradoxes, 2 vols, ed. by S. de Morgan, London 1872, 2. ed. 1915, repr. New York 1969; W. V. O. Quine, The Ways of Paradox and Other Essays, 2. ed. Cambridge, Mass.-London 1976, new ed. Cambridge, Mass. 1994; B. Russell, Les paradoxes de la logique, in: Revue de métaphysique et de morale 14, 1906, p. 627–650. Engl. transl. On 'Insolubilia' and their Solution by Symbolic Logic, in: B. Russell, Essays in Analysis, ed. by D. Lackey, London 1973.

4.6.6. L. Geldsetzer, ‚Sic et non' sive ‚Sic aut non'. La méthode des questions chez Abélard et la stratégie de la recherche, in : Pierre Abélard. Colloque international de Nantes (2001), ed. by J. Jolivet and H. Habrias, Rennes 2003, p. 407–415.

4.7.2. Weather forecasts and analysts' stock exchange prognoses are always "right" because they do not rule out that the weather or the stockmarket may develop contrary to their "probable" assertions. In contrast, hope in "probable" lottery winnings is always "wrong", which does not exclude a lucky draw by one or another participant.

4.7.5. Probability theories evade the answer to the question of what probability the half (or the 50 %) is a part of. Obviously what is meant is a half-truth and/or a half-falsity at the same time. We are told, at least, that the whole (or 100 %) of probability equals truth, and complete lack (or 0 %) of probability equals falsity! But don't these equivalences suggest that the "unit" (100 % = 1) of probability = truth, and the "absence" of this unit (0/100 % = 0) of probability = falsity? Which amounts to say: pure probability is truth, and: the unprobable is falsity.

4.8. On mathematical and logical aspects of probability calculation see R. Knerr, Goldmann Lexikon Mathematik, art. "Wahrscheinlichkeitsrechnung", Gütersloh-München 1999, p. 485–508; H. Rott, art. "Wahrscheinlichkeit", "Wahrscheinlichkeitsimplikation", "Wahrscheinlichkeitslogik", "Wahrscheinlichkeitstheorie" in: J. Mittelstrass, (ed.), Enzyklopädie Philosophie und Wissenschaftstheorie vol. 4, Stuttgart-Weimar 1996, p. 605–619 (with abundant bibliography) – See further C. Huygens, De ratiociniis in ludo aleae, in F. van Schooten, Exercitationes Mathematicae, Leiden 1657, p. 521–534; J. Bernoulli, Ars conjectandi, Basel 1713, repr. Bruxelles 1968; T. Bayes, An Essay Towards Solving a Problem in the Doctrine of Chances, 1763, new ed. in: Biometrica 45, 1958, p. 293–315; P. S. de Laplace, Théorie analytique des probabilités, Paris 1812, repr. Bruxelles 1967, Engl. transl. New York 1995; J. K. Fries, Versuch einer Kritik der Principien der Wahrscheinlichkeitsrechnung, Braunschweig 1842, also in: J. F. Fries: Sämtliche Schriften, ed. by G. König and L. Geldsetzer, vol. 14, Aalen 1974; J. Venn, The Logic of Chance, New York 1866, 4. ed. 1962; J. M. Keynes, A Treatise on Probability, London 1921, repr. New York 1979; R. v. Mises, Wahrscheinlichkeit, Statistik und Wahrheit. Einführung in die neue Wahrscheinlichkeitslehre, Wien 1928, 4. ed. 1972, Engl. transl. New York 1939 and 1981; A. N. Kolmogorov, Grundbegriffe der Wahrscheinlichkeitsrechnung. Ergebnisse der Mathematik (1933), repr. 1977, Engl. transl.: Foundations of the Theory of Probability, New York 1956; H. Reichenbach, Wahrscheinlichkeitslehre. Eine Untersuchung über die logischen und mathematischen Grundlagen der Wahrscheinlichkeitsrechnung, Leiden 1935, Engl. transl.: Berkeley-Los Angeles 1949 and 1971; A. Tarski, Wahrscheinlichkeitslehre und mehrwertige Logik, in: Erkenntnis 5, 1935/36, p. 174–175; R. Carnap, Logical Foundations of Probability, Chicago, Ill. 1950, 4. ed. London 1971; E. W. Adams, The Logic of Conditionals. An Application of Probability to Deductive Logic, Dordrecht-Boston 1975; K. R. Popper, A World of Propensities, Bristol 1990, Germ. transl. Tübingen 1995; – A. C. King and C. B. Read, Pathways to Probability. History of the

Mathematics of Certainty and Chance, New York 1963; T. L. Fine, Theories of Probability. An Examination of Foundations, New York 1973; E. J. Bitsakis and C. A. Nicolaides (eds): The Concept of Probability (Fundamental Theories of Physics 24), Dordrecht 1989; J. Hacking, The Emergence of Probability. A Philosophical Study of Early Ideas about Probability, Induction and Statistical Inference, London 1975, 2. ed. Cambridge-New York 1984; I. Schneider, ed., Die Entwicklung der Wahrscheinlichkeitstheorie von den Anfängen bis 1933, Einführungen und Texte, Darmstadt 1988, 2. ed. Berlin 1989; T. Hailperin, Probabilistic Logic in the twentieth Century, in: History and Philosophy of Logic 12, 1991, p. 71–110; – I. J. Good, Good Thinking. The Foundations of Probability and Its Applications, Minneapolis 1983; R. Chuaqui, Truth, Possibility, and Probability. New Logical Foundations of Probability and Statistical Inference, Amsterdam 1991; C. Howson, Theories of Probability, in: The British Journal for the Philosophy of Science 46, 1995, p. 1–32; J. E. Fenstad, Logic and Probability, in: E. Agazzi (ed.): Modern Logic. A Survey, Dordrecht 1981, p. 223–233; P. Milne, Can There Be a Realist Single-Case Interpretation of Probability, in: Erkenntnis 25, 1986, p. 129–132; M. Stevens, Do Large Probabilities Explain Better? in: Philosophy of Science 67, Suppl. vol. 2000, p. 366–390; T. A. F. Kuipers (ed.), What is Closer-to-the-Truth?, Amsterdam 1987; W. Lenzen, Glauben, Wissen und Wahrscheinlichkeit. Systeme der epistemischen Logik, Wien-New York 1980; G. Shafer and J. Pearl, eds., Readings in Uncertain Reasoning, San Meteo, Cal. 1990; S. Zwart, Approach to the Truth: Verisimilitude and Truthlikeness (Phil. Diss.), Groningen 1998.

4.8.11. Take for example the relation of known weather periods to the "whole" climate of larger time spans. Probable inferences from the known part or parts to the unknown whole seem highly (one can say 100 %) probable because we know that there exists weather at all times (which is itself a non-propositional complete induction of the concept "climate"). But there is no way of knowing (at least today) whether the characteristics of our present period are generic or specific with respect to the climate as a whole. It seems therefore that verification or falsification of a probable inference to the global climate remains true and false at once.

4.8.13. See L. A. Zadeh, Fuzzy Logic and Approximate Reasoning, in: Synthese 30, 1975, p. 407–428; T. T Balmer and M. Pinkal (eds), Approaching Vagueness. Amsterdam-New York-Oxford 1983; J. A. Goguen, The Logic of Inexact Concepts, in: Synthese 19, 1969, p. 325–373; D. McNeill and P. Freiberger, Fuzzy Logic, New York 1993.

5. In classical logic one distinguishes immediate and mediate inferences or conclusions. The former are understood as conjunctions of subject and predicate, contrapositions and subalternations. Aristotelian syllogisms, inductions and

analogies are presented as the prototypes of mediate inferences. Stoic inferences are usualy absent in the older textbooks. See, e.g., F. Ueberweg, System der Logik und Geschichte der logischen Lehren, 2. ed. Bonn 1865, §§ 82 ff. and §§ 99 ff, p. 206–361. – On modern inferential logic see: G. Gentzen, Untersuchungen über das logische Schließen, in: Mathematische Zeitschrift 39, 1934, p. 76–210 and 405–443; W. C. Salmon, The Foundation of Scientific Inference, Pittsburgh 1966; I. Niiniluoto and R. Tuomela, Theoretical Concepts and Hypothetico-Inductive Inference (Synthese Library 53), Dordrecht 1973; M. B. Hesse, The Structure of Scientific Inference, London 1974; E. Barnes, Inference to the Loveliest Explanation, in: Synthese 103, 1995, p. 251–302; P. Milne, Is there a Logic of Confirmation Transfer?, in: Erkenntnis 53, 2000, p. 309–335; G. Schurz, Normische Gesetzeshypothesen und die wissenschaftsphilosophische Bedeutung des nichtmonotonen Schließens, in: Journal for General Philosophy of Science / Zeitschrift für allgemeine Wissenschaftstheorie 32, 2001, p. 65–107; H. Leitgeb, Inference on the Low Level. An Investigation into Deduction, Nonmonotonic Reasoning, and the Philosophy of Cognition, Dordrecht 2004.

5.1. All of these simple or immediate implicative inferences cannot be distinguished from one another without pyramidal formalization. The customary notation of "whole" concepts by variables only allows us to formalize "if X then Y", which does not make visible the truth value of the linkage. This holds especially for correlative implications, which serve – applied to empirical causes and effects – as the sole means to logically formalize causal inferences. The logical condition of this application to empirical instances is that there exist a common generic concept identifying cause and effect as its dihairetic species. Obviously this logical condition could not be accounted for in the theories of causation of Aristotle, the Stoics and of Hume and Kant. – For the modern view see: J. C. Pitt and M. Tavel, Revolutions in Science and Refinements in the Analysis of Causation, in: Zeitschrift für allgemeine Wissenschaftstheorie / Journal for General Philosophy of Science 8, 1977, p. 48–62.

5.1.2. The "riddle of induction" or so-called Goodman's paradox was exhibited in Nelson Goodman's book "Fact, Fiction, Forecast", Cambridge, Mass. 1955, Germ. transl.: Tatsache, Fiktion, Voraussage, Frankfurt a. M. 1957. It plainly consists in the non-discrimination of material and correlative implication. – Many have attempted to propose a "solution", as: S. F. Barker and P. Achinstein, On the New Riddle of Induction, in: Philosophical Revue 69, 1960, p. 511–522; K. Eichner, Die Lösung des Goodman-Paradoxons: Goodmans Fehlschluß, in: Kant-Studien 66, 1975, p. 500–509; N. Stemmer, A Partial Solution to the Goodman Paradox, in: Philosophical Studies 34, 1976, p. 177–185 (see also 1.9. on induction).

5.3. For a comprehensive survey of Aristotle's logic including his syllogisms see I. M. Bochenski, Formale Logik, 3. ed. Freiburg-München 1970, p. 47–114, English translation: A History of Formal Logic, Notre Dame 1956, 2. ed. 1961; J. Lukasiewicz, Aristotle's Syllogistics from the Standpoint of Modern Formal Logic, Oxford 1951, 2. ed. 1957; G. Patzig, Die aristotelische

Syllogistik. Logisch-philologische Untersuchungen über das Buch A der 'Ersten Analytiken', Göttingen 1959, 3. ed. 1969, Engl. edition Dordrecht 1968; R. Smith, Logic, in: J. Barnes (ed.), The Cambridge Companion to Aristotle, Cambridge, Mass. 1995, p. 27–65; M. Malink, A reconstruction of Aristotle's modal syllogistic, in: History and Philosophy of Logic 27, 2006, p. 95–141.

5.3.2. For an overview and a critique of the singular syllogisms see L. Geldsetzer, Grundriß der pyramidalen Logik mit einer logischen Kritik der mathematischen Logik und Bibliographie der Logik, Internet HHU 2000, chapter 8 a. The linkages of the relevant concepts may be implemented in various ways, and they total 14 valid syllogisms, as Aristotle himself asserted. It is remarkable that there is to this day no consensus among logicians about the number of valid Aristotelian syllogisms in the range of the 256 syllogisms constructible by quantification of the subject concepts and the middle term.

5.3.3. See also L. Geldsetzer, Grundriß der pyramidalen Logik mit einer logischen Kritik der mathematischen Logik und Bibliographie der Logik, Internet HHU 2000, chapter 8 b.

5.5. On Stoic logic in general see B. Mates, Stoic Logic (Diss. Berkely), Los Angeles 1953; M. Mignucci, Il significato della logica stoica, Bologna 1965; M. Frede, Die stoische Logik (Abhandlungen der Akademie der Wissenschaften zu Göttingen), Göttingen 1974; Th. Ebert, Dialektiker und frühe Stoiker bei Sextus Empiricus. Untersuchungen zur Entstehung der Aussagenlogik (Hypomnemata 95), Göttingen 1991; Chrysippos of Soloi: Chrysippi Fragmenta logica et physica, ed. by H. von Arnim as vol. 2 of his Stoicorum Veterum Fragmenta, 4 vols, Leipzig 1903–1924, Leipzig 1903, 2. ed. Stuttgart 1964.

5.7. For the history and systematic of propositional logic see I. M. Bochenski, Formale Logik, 3. ed. Freiburg München 1970, p. 47–114, English translation: A History of Formal Logic, Notre Dame 1956, 2. ed. 1961; as also I. M. Bochenski and A. Menne, Grundriß der Logistik, 4. ed. Paderborn 1973, p. 27–66; J. Lukasiewicz, Zur Geschichte der Aussagenlogik, in Erkenntnis 5, 1935, p. 111–131. – Analytical philosophers take induction and deduction to be inferences, which they are not (see 1.8. ff). Following Wittgenstein they declare "elementary sentences" to be propositional instances which serve as premises for inferences to "general" propositions or "laws". Laws which allow exemptions are called "normic laws"! When empirical instances (which may also be statistical data) do confirm a law, they call it "monotonic inference"; if the instances do not, than the procedure is called "non-monotonic inference (or reasoning)". Deductions are held to be always "monotonic inferences". See G. Schurz, Normische Gesetzeshypothesen und die wissenschaftsphilosophische Bedeutung des nichtmonotonen Schließens, in: Journal for General Philosophy of Science / Zeitschrift für allgemeine Wissenschaftstheorie, 32, 2001, p. 65–107. – The huge debate between deductive Popperians and inductive Carnapians exposes the lack of conceptual reasoning and the overvaluation of "propositional logic" resulting from Wittgenstein's influence.

5.7.2. L.Wittgenstein, "Der Elementarsatz besteht aus Namen. Er ist ein Zusammenhang, eine Verkettung von Namen" (The elementary proposition consists of names. It is a connection, a concatenation of names), Tractatus logico-philosophicus 4.22., p. 49. What Wittgenstein had in mind was a conceptual expression linked together by expression-forming connectors, such as mathematical expressions like sums or products. Had he given a logical example he could only have instanced the sort of expressions customarily used as book titles, not assertive sentences.

5.8. See Nicholas of Kues' famous book De docta ignorantia / Die belehrte Unwissenheit (Of Learned Ignorance), ed. by P. Wilpert and H. G. Senger, 3 vols, 2. ed. Hamburg 1977–1979.

5.9. See J. Pearl, Probabilistic Reasoning in Intelligent Systems, Santa Mateo, Cal. 1988. – The classical foundations of mathematical probability were given by R. von Mises, Grundlagen der Wahrscheinlichkeitslehre, in: Mathematische Zeitschrift 5, 1919, p. 55–99; B. de Finetti, Foresight. Its Logical Laws, Its Subjective Sources, in: H. E. Kyburg and Smokler (eds), Studies in Subjective Probability, New York 1964, p. 95–158; J. M. Keynes, A Treatise on Probability, London 1957; K. E. Popper, The Propensity Interpretation of the Calculus of Probability, in: British Journal for the Philosophy of Science 10, 1959, p. 25–42; H. Jeffreys, Theory of probability, 3. ed. Oxford 1961; R. Carnap, Logical Foundations of Probability, Chicago 1951; H. E. Kyburg, Probability and the Logic of Rational Belief, Middletown 1961. – The classroom textbook in Germany is W. Stegmüller, Probleme und Resultate der Wissenschaftstheorie und Analytischen Philosophie, Vol. IV: Personelle und statistische Wahrscheinlichkeit, Teil 1: Personelle Wahrscheinlichkeit und rationale Entscheidung; Teil 2: Statistisches Schließen – Statistische Begründung – Statistische Analyse, (Personal and Statistical Probability, part I: Personal Probability and Rational Decision; part II: Statistical Conclusion – Statistical Justification –Statistical Analysis), Berlin 1973. See my review in Philosophy and History 11, 1978, p. 42–48.

5.9.1. H. Rott, "Wahrscheinlichkeitslogik" (Logic of probability), in J. Mittelstrass (ed.), Encyklopädie Philosophie und Wissenschaftstheorie, vol. 4, p. 612, Stuttgart 1996.

5.9.3. See H. Blanchard, The Case for Determinism, in: S. Hook (ed.), Determinism and Freedom in the Age of Modern Science, New York 1963; J. Earman: A Primer on Determinism, Dordrecht 1986; J. Gleick, Chaos: Making a New Science, New York 1987. – Causal connection between parts of the world has been assumed rather than deduced in occidental science since the Pre-Socratics. Aristotle notably posited four kinds of causes, namely formal, material, final and efficient, and the Stoics reduced these to the latter two. Progress in modern science has been mainly considered to consist in the elimination of final causes, so that only efficient causes remained – at least in the exact sciences. The principal critiques of these assumptions concerning causes are those of Nagarjuna (see 2.5.2.) in India and China and of Sextus Empiricus (see 2.4.4.) in occidental antiquity, but obviously, neither was ever taken seriously. Nagarjuna (ca. 2.–3. century A. D, see my translation and commentary to his Madhyamaka-karika, Hamburg 2010) replaced causality with "pratitya samutpada" which means that all things in appearance are intervowen or interdependent and contain no substantial kernels fit to be distinguished as causes and effects. Sextus hinted at the fact that so-called causes are no longer present and therefore only recollected ideas when an effect is empirically found to be present, and similarly prognoses are imagined effects of actual data declared to be causes. Both positions fit quite well with modern constitutions of causality by Hume ("post hoc"-thesis, which presupposes temporal consciousness) and Kant (categorical supposition in transcendental consciousness).

6.1. See W. Stegmüller, Probleme und Resultate der Wissenschaftstheorie und Analytischen Philosophie, Vol. 2: Theorie und Erfahrung, 2nd part: Theorienstrukturen und Theoriendynamik (Problems and Results of the Philosophy of Science and of the Analytical Philosophy, vol. 2: Theory and Experience, 2nd part: Theory Structure and Dynamics of Theories), Berlin-Heidelberg-New York 1973, especially Chapter VIII: The Structure of ripe physical theories after Sneed, p. 27–152; W. Balzer and M. Heidelberger (eds), Zur Logik empirischer Theorien, Berlin-New York 1983; M. Carrier, The Completeness of Scientific Theories (The University of Western Ontario Series in Philosophy of Science 53), Dordrecht 1994; J. W. McAllister, The Simplicity of Theories. Its Degree and Form, in: Journal for General Philosophy of Science / Zeitschrift für allgemeine Wissenschaftstheorie 22 1991, p. 1-14; R. E. Grandy, Theories of Theories. A View from Cognitive Science, in: J. Earman (ed.), Inference, Explanation, and Other Frustrations. Essays in the Philosophy of Science, Berkeley, Cal. 1992, p. 216–233.

6.2. All well-definied "hard core" concepts of a theory can be arranged and ordered in a conceptual pyramid. This kind of logical formalization allows us to read off the definitions from the positions of the concepts in the pyramid, and the resulting linkages between them as true, false or true-false propositions. Examples of such formalizations are the number-theory (see 1.6.) and the theory of connectors (see 2.7.) in this book. Further examples are exhibited in L. Geldsetzer, Über das logische Prozedere in Hegels Phänomenologie des Geistes (On the logical procedure in Hegel's Phenomenology of Spirit), in: Jahrbuch für Hegelforschung 1, 1995, p. 43–80 and L. Geldsetzer. Dao als metaphysisches Prinzip bei Lao Zi, in: Monumenta Serica, Journal of Oriental Studies 47, 1999, p. 237–254. – Pyramidal formalization is well suited to systematically assemble the "logical parts" of a theory, to distinguish these from mere verbiage used to "clothe" them, and to exhibit the essence of the theory on a single page.

6.3. See Euclid: The Thirteen Books of Euclid's, 3 Vols, Cambridge 1905–1925, repr. New York 1956, German transl. by J. F. Lorenz and ed. by C. B. Mollweide, 5.ed. Halle 1824; Edition by E. S. Stamatis: Die Elemente Buch I – XIII, Leipzig 1933–1937, repr. Darmstadt 1962; Latin transl. by J. L. Heiberg and H. Mengs in Euclidis Opera Omnia, 8 vols and suppl., Leipzig 1893–1916, repr. Leipzig 1969–1977.

6.3.1. R. Carnap tried to resolve a Kantian problem, namely to coordinate highest "theoretical" (or: "unanschauliche") concepts or categories with empirical "sensory" ("anschauliche") concepts. See R. Carnap, Logische Syntax der Sprache, Wien 1934, 2. ed. Wien-New York 1968; Engl. edition London-New York 1937, new transl. by A. Smeaton, London N. J. 1959, 6. ed. 1964. This presupposes that "theoretical concepts" are – in Kantian understanding – products of pure thinking, and that empirical concepts are the product of sensory perception, so that the two kinds have nothing in common. Kant's authority prevented his critics from realizing that (to use Kant's own words) abstract theoretical concepts would therefore be "empty", and concepts formed on the basis of sensory perception would be "blind". It is noteworthy that Kant invoked "imagination" to give a solution. And so did Carnap, continuing the Vienna Circle's search for a criterion of meaning of general concepts by introducing "bridge concepts" to mediate between the two kinds. Obviously this is the result of a half-hearted empiricism which sets apart logic and mathematics as "pure theoretical or formal language", untouched by sensory contamination. In contrast, I show in this book that theoretical concepts can and must be inductively derived from contentful empirical perceptions and only then can serve as defined axiomatic concepts for deductions.

6.3.2. See M. Carrier, art. "Theoriebeladenheit" in: J. Mittelstraß (ed), Enzyklopädie Philosophie und Wissenschaftstheorie, vol. 4, Stuttgart-Weimar 1996, p. 272–274.

6.3.3. The absence of this insight is a regrettable deficiency of Berkeley's metaphysical theory. It would have allowed him to explain clearly what his critics did in fact see and affirm about infinitesimal concepts (beyond the "minimum sensibile"). In occidental metaphysics and ontology, however, there has never been an induction of "nothing". But one can learn from examples of this venture in Eastern philosophies, as for example in the Lao Zi Dao De Jing. See my "New philosophical translation from the Chinese" in: Internet HHU Düsseldorf; also in: Asiatische Philosophie. Indien und China. CD-Rom, Digitale Bibliothek (Directmedia Publishing GmbH), Berlin 2005. An other example is Nagarjuna in his Mulamadhyamaka-Karika; see: Nagarjuna, Die Lehre von der Mitte. Mula-madhyamaka-karika / Zhong Lun, Chinesisch-Deutsch, trad. and commentary by L. Geldsetzer, Hamburg 2010, p. 120 and p. 133–138.

6.3.7. That realism is a dialectical expansion of idealism is paradigmatically demonstrated in Kant's "Critique of Pure Reason". Its first edition of 1781 was conceived as an idealistic theory, as can be seen mainly in its original introduction. In response to the review of Christian Garve, (published in the edition of Kant's Prolegomena by K. Vorländer, Hamburg 1951, p. 167–174) who hinted at the proximity to Berkeley's idealism, its second edition of 1787 was obviously reconstructed as a realistic theory, as is declared in its new introduction and some insertions in the text (esp. "Refutation of idealism") and the suppression of the former introduction. Both versions are now distinguished as versions A and B and printed side by side. The bulk of the text of the "Critique" remained the same and was interpreted in an idealistic sense by German idealists like K. L. Reinhold, J. G. Fichte, F. W. J. Schelling, G. F. W. Hegel and A. Schopenhauer, and in the realistic sense by German realists like J. F. Fries, J. F. Herbart, Ed. Beneke and their followers, including many analytical philosophers. See K Vorländer, Geschichte der Philosophie vol. III/1 (Philosophy in the First Half of the nineteenth Century), revised and enlarged by L. Geldsetzer, Hamburg 1975, p. 23–160.

6.4.1. The "credo quia absurdum" is a characterization of Tertullian's dialectical theses against "logical" heretics, as for example: "Crucifixus est dei filius; non pudet, quia pudendum est. Et mortuus est Dei filius; prorsus credibile est, quia ineptum est. Et sepultus ressurexit; certum est, quia impossibile est" (The son of God was crucified; that is no shame because it is a shame. And the son of God was dead; this is almost uncredible because it is stupid. And after having been burried he was ressurected; this is certain because it is impossible" (Tertullian, "De carne Christi"), cit. from F. Ueberweg, Grundriß der Geschichte der Philosophie, vol. 2, 15. ed. Basel-Stuttgart 1956, p. 47. – P. Abélard in his work "Sic et Non" (in "Petri Abaelardi Opera Omnia", contained in Patrologia Latina vol. 178, ed. by J. P. Migne, repr. Tournholt s. a.) showed that the main articles of belief of the sacred scriptures and the saints, in contrast to non-dogmatic assertions, were all formulated as contradictory assertions. See L. Geldsetzer, 'Sic et non' sive 'Sic aut non'. La méthode des questions chez Abélard et la stratégie de la recherche, in: J. Jolivet and H. Habrias (eds), Pierre Abélard. Colloque international de Nantes, Rennes 2003, p. 407–415. – Nicholas of Kues in his work Apologia Doctae Ignorantiae (1449) remarks in good Platonic humor: "Cum nunc Aristotelica secta praevaleat, quae haeresin putat esse oppositorum coincidentiam, in cuius admissione est initium ascensus in mysticam Theologiam, in ea secta nutritis haec via penitus insipida, quasi propositi contraria, ab eis procul pellitur, ut sit miraculo simile – sicuti sectae mutatio – reiecto Aristotele eos altius transilire" / Since today the Aristotelian sect prevails which considers the coincidence of the opposities (= coincidence of the contradictories) a heresy, although this admission is the very beginning of the ascent to mystical theology, to the scholars of this sect this way appears as obviously stupid and contrary to their own ends, so they reject it totaly. It would appear a miracle – like a revolution of the sect – if they would reject Aristotle and progress to higher insights. Apologia doctae ignorantiae / Verteidigung der wissenden Unwissenheit, in: Nikolaus von Kues, Philosophisch-Theologische Schriften, ed. by L. Gabriel, vol. I, Wien 1964, p. 530–531. – G. W. Leibniz, transfering this "mystical ascent" into the construction of the differential calculus, writes in his "Generales inquisitiones de analysi notionum et veritatum" (§ 66) of the year 1686: "(Wenn) die Differenz zwischen dem, was zusammenfallen (d. i. identisch werden) soll, kleiner wird als jede beliebige vorgegebene Größe, so ist bewiesen, daß der betreffende Satz wahr ist" / "(if) the difference between that which should coincide (that is, become identical) becomes smaller than any given quantity, it is proved that the relevant sentence is true"; cit. from O. Becker, Grundlagen der Mathematik in geschichtlicher Entwicklung, 2. ed. Freiburg-München 1964, p. 360. The differential calculus is one instance of Leibniz's "law of continuity", "kraft dessen man die Ruhe als eine unendlichkleine Bewegung – d. h. als äquivalent einer Unterart ihres Gegenteils – ansehen kann, das Zusammenfallen zweier Punkte als eine unendlichkleine Entfernung zwischen ihnen, die Gleichheit als Grenzfall der Ungleichheit usw." / "on the basis of which one can consider stillstand as an infinitesimally small motion – that is equivalent to a subspecies of its contrary – coincidence of two points as an infinitesimally small distance between them, equality as a boundary case of inequality, etc.", in his Letter to Varignon, Febr. 2, 1702, cit. from O. Becker, p. 167.

6.4.2. Ralf Goeres, "Die Entwicklung der Philosophie Ludwig Wittgensteins unter besonderer Beruecksichtigung seiner Logikkonzeptionen" (The Development of Wittgenstein's Philosophy with Special Reference to his Conceptions of Logic), Wuerzburg, 2000, p. 355. – L. Geldsetzer, "Ueber das Logische Prozedere in Hegels Phaenomenologie des Geistes" (Concerning Hegel's Logical Procedure in the Phenomonology of Spirit), in: Jahrbuch für Hegelforschung, ed. by H. Schneider, Vol. I, Sankt Augustin 1995, pp. 43–80, particularly p. 75 and p. 80. The pyramid of Hegel's principal concepts in the "Phenomenology" is also exhibited in L. Geldsetzer, Grundriß der pyramidalen Logik mit einer logischen Kritik der mathematischen Logik und Bibliographie der Logik, Internet Heinrich-Heine-University 2000, Appendix.

6.5. The state of the art of theory building may be reviewed in D. A. Anapolitanos, Theories and their Models, in: Zeitschrift für allgemeine Wissenschaftstheorie / Journal for General Philosophy of Science 20, 1989, p. 201–211; M. Carrier, The Completeness of Scientific Theories (The University of Western Ontario Series in Philosophy of Science 53), Dordrecht 1994; R. E. Grandy, Theories of Theories. A View from Cognitive Science, in: J. Earman (ed.): Inference, Explanation, and Other Frustrations. Essays in the Philosophy of Science, Berkeley, Cal. 1992, p. 216–233; J. W. McAllister, The Simplicity of Theories. Its Degree and Form, in: Journal for General Philosophy of Science / Zeitschrift für allgemeine Wissenschaftstheorie 22, 1991, p. 1–14; H. Radermacher, Der Begriff der Theorie in der kantischen und analytischen Philosophie, in: Zeitschrift für allgemeine Wissenschaftstheorie / Journal for General Philosophy of Science 8, 1977, p. 63–76; E. Scheibe, Two Types of Successor Relations between Theories, in: Zeitschrift für allgemeine Wissenschaftstheorie Journal for General Philosophy of Science 14, 1983, p. 68–80; F. Suppe, Theories. Their Formation and Their Operational Imperative, in: Synthese 25, 1972, p. 129–164; F. Suppe (ed.), The Structure of Scientific Theories, Urbana-Chicago-London 1974, 2. ed. 1977; F. Suppe, Understanding Scientific Theories. An Assessment of Developments 1969–1998, in: Philosophy of Science 67, Suppl. vol. 2000, p. 102–115; J. S. Wilkins, The Evolutionary Structure of Scientific Theories, in: Biology and Philosophy 13, 1998, p. 479–504.

6.5.1. See L. Geldsetzer and F. Rotter (eds), Der Methoden- und Theorienpluralismus in den Wissenschaften (Studien zur Wissenschaftstheorie 6, ed. by A. Diemer), Meisenheim 1971.

6.5.2. One of the best overviews of mathematics from the philosophical point of view is still Charles Parsons' article "Mathematics, Foundations of", in: P. Edwards (ed.), The Encyclopedia of Philosophy, vol. V, London-New York 1967, p. 188–212.

6.5.3. In Newton's time Ruggiero Giuseppe Boscovich's physical system "Philosophiae naturalis theoria redacta ad unicam legem virium in natura existentium", Vienna 1758 and Venice 1763, new Engl.-Latin ed. "A Theory of Natural Philosophy", Chicago-London 1922, exhibited an alternative concept of atoms and anticipated some ideas of the relativity theory, but was never recognized as a competitor to Newton. The same holds for many critiques of relativity theory and quantum mechanics which, notwithstanding the contradictions and paradoxes (now called "decoherences") of these theories, have seldom any chance of getting published in the relevant organs of the physics community.

6.6. See also U. Charpa, Philosophische Wissenschaftshistorie. Grundsatzfragen / Verlaufsmodelle, Braunschweig-Wiesbaden 1995, p. 212–221 on "Nontranslatable Concepts and Theories" and "Indefinite data".

6.6.1 Nota bene: there may be longstanding discussions among researchers whether a "phenomenon" is an empirically observed fact or an artificially introduced non-fact (due, for example to the experimental apparatus). Artifacts have hitherto not found adequate consideration, as the absence of relevant articles in the representative dictionaries shows.

6.6.2. U. Charpa, op. cit. p. 214 on the received view of astronomical classifications at that time.

6.6.3. See W. Stegmüller, Probleme und Resultate der Wissenschaftstheorie und Analytischen Philosophie, Vol. 2: Theorie und Erfahrung, 2nd part: Theorienstrukturen und Theoriendynamik (Problems and Results of the Philosophy of Science and of the Analytical Philosophy, vol. 2: Theory and Experience, 2nd part: Theory Structure and Dynamics of Theories), Berlin-Heidelberg-New York 1973, especially chapter IX on "Theory dynamics", p. 153–311, on the debate over T. S. Kuhn's "The Structure of Scientific Revolutions" (1966), 2. ed Chicago 1970; also W. Stegmüller, The Structure and Dynamics of Theories, Berlin 1976. – The debate shows that hermeneutics suffers from the same error which is current in the thesis that knowledge is per se true knowledge and false knowledge is not knowledge at all. Interpreters likewise suppose that understanding is per se true understanding and false understanding is not interpretation at all. There hasn't been sufficient attention given to the fact that there is also true understanding of false assertions and theories as well as false understanding of true assertions and theories. See also L. Geldsetzer, Truth, Falsity and Verisimilitude in Hermeneutics (Contribution to the Symposium Analytical Philosophy and Philosophy of Science, July 23–25, 1996 in Beijing), in: Internet HHU Düsseldorf 1998.

6.6.5. Professional scientists' interest in and know-how concerning the production of good disciplinary bibliographies is currently in full decline. Their role has been assumed by so-called citation indexes, which develop their own canonical network of bibliographies of the cited texts and exclude everything unfitting from the notice of the professional community.

7. On the state of the art of axiomatics see H. Pulte, Axiomatik und Empirie. Eine wissenschaftstheoriegeschichtliche Untersuchung zur Mathematischen Naturphilosophie von Newton bis Neumann, Darmstadt 2005; H. Schüling, Die Geschichte der axiomatischen Methode im 16. und beginnenden 17. Jahrhundert. Wandlungen der Wissenschaftsauffassung, Hildesheim-New York 1969; F. Suppe, Axiomatization, in: A Companion to the Philosophy of Science, ed. by W. H. Newton Smith, Oxford 2001, p. 9–11. – The main promoter of axiomaticism as "proof theory" in mathematics and mathematical logic was David Hilbert. His philosophy was honored by the denomination "Formalism" *tout court.* See V. Peckhaus, Hilberts Logik. Von der Axiomatik zur Beweistheorie, in: NTM (International Journal of History and Ethics of Natural Sciences, Technology and Medicine, New Series) 3, 1995, p. 65–86. Hilbert declared in the famous manifesto of his axiomaticism: "Everything that can be an object of scientific thought at all, as soon as it is ripe for theory construction, becomes subject to the axiomatic method and thereby indirectly to mathematics", in: D. Hilbert, Axiomatisches Denken, in: Mathematische Annalen 78, 1918, p. 405–415, repr. in: Gesammelte Abhandlungen III, Berlin 1935, p. 146–156; reprinted also New York 1965. However, since his axioms were defined as "forms for propositions" and concepts inserted into them as "implicitely defined" and thus open for any interpretation by

models, the customary presuppositons of axioms in deductive proofs is rather a "confession of belief" or of the convictions common to the members of a special scientific community than the formalization of understandable propositions and clear and distinct concepts. This was clearly observed long ago by Nicholas Cusanus in his "De Docta Ignorantia" where he says: "In omni enim facultate quaedam praesupponunter ut principia prima, quae sola fide apprehenduntur, ex quibus intelligentia tractandorum elicitur" / In every region of science some propositions are presupposed as principles (axioms) which are only comprehensible through belief and out of which the knowledge of the object of research is developed; see Nikolaus von Kues, De Docta Ignorantia / Belehrte Unwissenheit (1440), vol. III, ed. by H. G. Senger, Hamburg 1977, p. 74–75.

7.1. They were introduced by Aristotle as axioms in the sense of the most general propositional principles of logic and have been maintained to this day in textbooks of logic and mathematics. And Aristotle's defense of them as axioms has remained the standard of professional "justification" in the disciplines: "Some, indeed, demand to have the law (scl. of contradiction) proved, but this is because they lack education; for it shows lack of education not to know of what we should require proof, and of what we should not. For it is quite impossible that everything should have a proof; the process would go on to infinity, so that even there would be no proof." (Metaphysics book IV, 2, 1006a), see Aristotle, The Metaphysics, transl. by H. Tredennick, Cambridge, Mass. 1956, p. 163. – G. W. Leibniz and Chr. Wolff attempted to add a fourth "principle of sufficient reason" (principium rationis sufficientis) for every logical proceeding and for proofs. See G. W. Leibniz, Specimen inventorum de admirandis naturae generalis arcanis, in: Die Philosophischen Schriften, ed. by C. I. Gerhardt, vol. VII, Berlin 1890, p. 309; Chr. Wolff, Philosophia prima sive ontologia, Frankfurt -Leipzig 1730, 2nd ed. 1736, §§ 67–70. This principle was never accepted as a fourth logical axiom. It did, however, establish itself as a powerful demand that all scientific argumentation should be pursued all the way back to axioms (as so-called necessary conditions) or at least to accepted theorems (as sufficient conditions).

7.1.2. For current considerations about the principle of the third see F. von Kutschera, Der Satz vom ausgeschlossenen Dritten. Untersuchungen über die Grundlagen der Logik, Berlin-New York 1985; N. Vasallo, Sulla problematicità del principio del terzo escluso. Linguisticità e senso concreto del principio nella lettura intuizionista di L. E. J. Brouwer, in: Epistemologia 23, 2000, p. 99–118.

7.1.5. See G. W. Leibniz, "Specimen calculi universalis" in: Die Philosophischen Schriften, ed. by C. I. Gerhardt, vol. VII, Berlin 1890, p. 219. He writes: "Eadem sunt quorum unum in alterius locum substitui possit, salva veritate" / "Things are the same if one of them can be substituted in the place of the other while preserving truth". – Nota bene: it is only Leibniz' formalization of his principle that is inadequate. Its practical value lies in the fact that the "indiscernible" cannot be distinquished from nor substituted for itself.

7.2. On these criteria for axioms see e.g. R. Knerr, Goldmann Lexikon Mathematik, Gütersloh-München 1999, p. 34 f. – That axioms have to be evident or "categorical" (as the new denomination goes) is rather a heritage of Platonic "innate ideas" and Stoic "common notions". Evidence has commonly been replaced by "intuition" or simply by "belief". – See also: R. Schantz, Was ist sinnliche Evidenz? in: Logos, Neue Folge 5, 1998; W. Stegmüller, Der Evidenzbegriff in der formalisierten Logik und Mathematik, in Wiener Zeitschrift für Philosophie, Psychologie und Pädagogik 4/4, 1953, p. 288–295; G. Shafer, A Mathematical Theory of Evidence, Princeton 1976.

7.2.1. See K. Gödel, "Über formal unentscheidbare Sätze der Principia Mathematica und verwandter Systeme", in Monatshefte für Mathematik und Physik 38, 1931, p. 173–198, Engl. tr. "On Formally Undecidable Propositions of Principia Mathematica and Related Systems", Princeton 1934, reprinted 1962. See W. Stegmüller, Unvollständigkeit und Unentscheidbarkeit. Die metamathematischen Resultate von Gödel, Church, Kleene, Rosser und ihre erkenntnistheoretische Bedeutung, 3rd ed. Wien-New York 1973; I. Grattan-Guiness, The Search for Mathematical Roots 1879–1940. Logics, Set Theories and the Foundations of Mathematics from Cantor through Russell to Gödel, Princeton-Oxford 2000. – As generally understood by mathematicians, this and some other proofs of undecidability, contradiction and non-completeness dealt fatal blows to mathematical axiomatics and the so called Hilbert programm of axiomatization as also to logicism in toto. See St. Donati, I fondamenti della matematica nel logicismo di Bertrand Russell, Firenze s. a. (2004), chapter 7 "On the logicism of Russell". p. 861–944, esp. on Quine's and Gödel's critique, p. 908–927 and p. 927–944. What was in fact demonstrated was rather that the problem of axioms cannot be treated without taking account of the dialectical character of some mathematical and logical axiomatic concepts.

7.4. These examples of the real logical axioms show that they are not at all independent from one another. "Truth = not-falsity" and "falsity = not-truth" define each other. "Truth is not falsity" and "falsity is not truth" are axiomatic propositions.

7.4.1. "Coherence" as the intension of the concept of truth is inductively established from that which true logical elements, such as propositions and inferences, have in common: namely the connection of regular concepts by proposition-forming connectors used in accordance with their proper (truth-conducive) definitions. So one reads off in the pyramidal formalism that "AB is A" and "AB is not AC" are true propositions. – Traditionally, coherence (or consistency) in logic is understood as the character of a system of propositions without any contradiction between them. In this sense it is also identified with the freedom from contradictions exhibited by some theories. But this explains truth only as the counterpart of "contradiction as falsity", which is a false explanation, because contradiction is true and false at once. Hence coherence requires a proper inductive explanation, as here proposed. Regarding the problems involved see also: T. Shogenji, Why does coherence appear truth-conducive?, in: Synthese 157, 2007, p. 361–372, and N. Pfeifer and G. D. Kleiter, Coherence and Non-monotonicity in Human Reasoning, in: Synthese 145, 2005, p. 93–101. – Coherence is usually linked with idealistic epistemology, which is also the background of the present book. Realist epistemologists, however, maintain that coherence is not "truth-conducive". See: D. Lewis, Forget about the 'Coherence Theory of Truth', in: Analysis 61, 2001, p. 275–280; and E. Olson, Why Coherence is not Truth-conducive, in: Analysis 61, 2001, p. 236–241. Realists cherish the so-called correspondence theory of truth, which goes back to Aristotle's and Thomas Aquinas' definition of correspondence as "adaequatio rei et intellectus". For them this "adequation of reality and forms of thinking" is expounded in semantical interpretations of otherwise empty syntactical formalisms. But this definition shares the dialectical structure of realism. "Adequation" has here the double meaning of identifying "things and thinking consciousness" (as does idealism) and at the same time distinguishing them, which certainly is a "mission impossible". To be clear: thinking without content is not thinking at all, and any thing or reality not consciously experienced and logically retrieved is mere hallucination.

7.4.2. To define pure falsity in logic is quite unusual because of the age-old prejudice that logical falsity lies in contradiction or rather is identical with it. Therefore there is need to define falsity inductively in order also to show what it is in true-false (or probable) propositions, that is, contradictory ones. Logical falsity is what false propositions and inferences have in common, namely the connection of regular concepts by proposition-forming connectors in ways not defined as truth-conducive. So one easily reads off in pyramidal formalization that "AB is AC" or that "AB is not A" are false propositions and not at all contradictory ones! One can also read off that "If AB than AC" is a false material and at the same time a false formal implicative proposition, but a true correlative implication. Unless the three distinct implicative connections are distinguished in the pyramid the proposition appears both true and false at once, which in turn results in one of the well-known "riddles of implication"! – Recall Wittgenstein's postulated ideal that one should "see" the truth value of propositions in their formalization. – There exists considerable literature on fallacy, falsification, fallibility and fallibilism in logic and epistemology, but "falsity" is almost absent there. And no article on "falsity" is found in the representative dictionaries.

7.5. Contradiction comprises both fused dihairetical (dialectical) concepts, which have no truth value, as well as true-false or probable propositions and inferences. It is not definable by induction from these elements, because contradictory concepts are neither true nor false, and true-false propositions allow only inductions of either truth or falsity. Therefore contradiction must be defined by deduction from truth and falsity as their fused unity. Obvious propositional contradictions as for example "AB is A and not-A" are easily visible as true and false at once in pyramidal formalization. But contradictory concepts are not easily detectable in practise, and traditional formalisms are not apt to formalize them at all. And this notwithstanding the fact that they are commonly used in contradictory propositions (especially in paradoxes). For example, B. Russell's "class of all classes" (or "set of all sets") is still understood as a regular concept in mathematical logic and mathematics. And so are "possibility" in modal logics, and probability or verisimilitude in logics and mathematics. – See also: H. A. Zwergel, Principium contradictionis. Die aristotelische Begründung des Prinzips vom zu vermeidenden Widerspruch und die Einheit der Ersten Philosophie, Meisenheim 1972; M. Wolff, Der Begriff des Widerspruchs. Eine Studie zur Dialektik Kants und Hegels, Königstein 1981; L. Geldsetzer, Über das logische Procedere in Hegels 'Phänomenologie des Geistes'. In: Jahrbuch für Hegelforschung, ed. by H. Schneider, vol. 1, St. Augustin 1995, p. 43–80; V. Raspa, In-contraddizione. Il principio di contraddizione alle origini della nuova logica, Trieste 1999; F. G. Asenjo, A Calculus of Antinomics, in: Notre Dame Journal of Formal Logic 7, 1966, p. 103–105; S. Jaskonski, Propositional Calculus for Contradictory Deductive Systems (Polish 1948), in: Studia Logica 24, 1969, p. 143–157. – G. Priest attracted much attention with his thesis in "In Contradiction. A Study of the Transconsistent", Dordrecht 1987, 2nd ed. 2006, that contradictions can unite two "true" components. Since then Priest and his followers have developed this idea into a "Paraconsistent logic" or as it is also called "Dialetheism". See: M. Bremer, Wahre Widersprüche. Einführung in die parakonsistente Logik, Sankt Augustin 1998. But this was an error which Kant had already committed in his "Dialectic of pure reason" concerning the "dynamical antinomies" (as distinct from the "mathematical antinomies" which he declared false in both parts, see: Prolegomena § 53 ff.). Obviously this is alltogether "Beyond the Limits of Thought", as another of Priest's books (Cambridge 1995) asserts.

7.6. It is usually thought that logic has mainly to do with truth and falsity and their distinction. But logic and especially mathematical logic and mathematics are also concerned with definitions and the well-formedness of concepts and expressions. And this explains the stupendous flourishing of specialized logical disciplines. Currently one can count more then a hundred denominations of specialized logics. See L. Geldsetzer, Bibliography of logic and logical foundations of mathematics up to 2008, Internet HHU Duesseldorf 2008.

7.7. See Platon's dialogue "Ion" (especially 536 ff.) where Plato demonstrates that poets do not command adequate knowledge of the reality about which they devise fables; and "Politeia" (377b ff. and 595a ff.), where he recommends expelling the poets and rhapsodists from the Polis because they produce imitations of phenomena in order to arouse the lower sensual appetites of the citizens and do not command scientific knowledge of the "ideas", which alone guarantee truth. – In contrast, Aristotle (in his "Peri poeseos" / On Poetry) underlines the "cathartic" character of the arts including poetry and maintains that poetry is "more philosophical and also more important than historiography (scl. as empirical knowledge!). For poetry speaks rather about the general, whereas historiography speaks about the particulars" (On Poetry 1451a36) – See Wl. Tatarkiewicz, Geschichte der Ästhetik, vol. I, Basel-Stuttgart 1979, p. 139–167 (on Plato's esthetics) and p. 167–200 (on Aristotle's esthetics).

7.8. See the classic work of H. Vaihinger, Die Philosophie des Als -Ob. System der theoretischen, praktischen und religiösen Fiktionen der Menschheit auf Grund eines idealistischen Positivismus, 10. ed. Leipzig 1927; further: D. Koriako, Unerweisliche Sätze, erdichtete Begriffe. Kant über den Gebrauch mathematischer Argumente in der Philosophie, in: Studia Leibnitiana 30, 1998, p. 24–48; F. Crahay, Le formalisme logico-mathématique et le problème du non-sens, Paris 1957; O. Weinberger, Faktentranszendente Argumentation, in: Zeitschrift für allgemeine Wissenschaftstheorie / Journal for General Philosophy of Science 6, 1975, p. 235–251 (the author includes logic itself into this realm of the thinkable); M. J. Wreen, Most Assur'd of What He Is Most Ignorant, in: Erkenntnis 44, 1996, p. 341–368; P. R. Gross and N. Levitt, Higher Superstition. The Academic Left and its Quarrels with Science, Baltimore 1994; A. Lugg, Pseudoscience as Nonsense, in: Methodology and Science 25, 1992, p. 91–101; H. C. D. G. De Regt, To Believe in Belief. Popper and van Fraassen on scientific realism, in: Journal for General Philosophy of Science / Zeitschrift für allgemeine Wissenschaftstheorie 37, 2006, p. 21–39 (the author refutes realistic belief in unobservables but defends this belief on pragmatic grounds); K. L. Pfeiffer, Zum systematischen Stand der Fiktionstheorie, in: Journal for General Philosophy of Science / Zeitschrift für allgemeine Wissenschaftstheorie 21, 1990, p.135–156.

7.8.1. See M. A. Bishop, Why Thought Experiments are not Arguments, in: Philosophy of Science 66, 1999, p. 534–541; J. R. Brown, The Laboratory of the Mind. Thought Experiments in the Natural Sciences, London 1991; D. Cohnitz, Gedankenexperimente in der Philosophie, Paderborn 2006; D. Cole, Thought and Thought Experiments, in: Philosophical Studies 45, 1984, p. 431–444; S. Häggqvist, Thought Experiments in Philosophy, Stockholm 1996; T. Horowitz and G. J. Massey (eds), Thought Experiments in Science and Philosophy, Savage, Maryland 1991; J. W. McAllister, The Evidential Significance of Thought Experiment in Science, in: Studies in History and Philosophy of Science 27, 1996, p. 233–250; M. Bunzl, The Logic of Thought Experiments, in: Synthese 106, 1996, p. 227–240. See also the debate in the Journal for General Philosophy of Science 34–37, 2003–2006.

7.9. Concerning a not-so-new scientific phenomenon, see: K.-F. Wessel and M. Koch, Lügen ist überhaupt das Kennzeichen unserer Zeit. Über einen unveröffentlichten Briefwechsel zwischen Max Born und Friedrich Herneck, in: Berichte zur Wissenschaftsgeschichte 18, 1995, p. 27–33; W. Broad, and N. Wade, Betrug und Täuschung in der Wissenschaft, Basel 1983; A. Lugg, Bunkum, Flim-Flam and Quackery. Pseudoscience as a Philosophical Problem, in: Dialectica 41, 1987, p. 221–230; M. C. LaFollette, Stealing into Print. Fraud, Plagiarism, and Misconduct in Scientific Publishing, Berkeley 1992.

Name Index

L. Geldsetzer and R.L. Schwartz, *Logical Thinking in the Pyramidal Schema of Concepts: The Logical and Mathematical Elements*, DOI 10.1007/978-94-007-5301-3,

W

Z

Subject Index

L. Geldsetzer and R.L. Schwartz, *Logical Thinking in the Pyramidal Schema of Concepts: The Logical and Mathematical Elements*, DOI 10.1007/978-94-007-5301-3,

D

E

MIX
Papier aus verantwortungsvollen Quellen
Paper from responsible sources
FSC® C105338

If you have any concerns about our products,
you can contact us on
ProductSafety@springernature.com

In case Publisher is established outside the EU,
the EU authorized representative is:
Springer Nature Customer Service Center GmbH
Europaplatz 3, 69115 Heidelberg, Germany

Printed by Libri Plureos GmbH
in Hamburg, Germany